BOOK XVII IN THE RAIDING FORCES SERIES

THE MAGNIFICENT MISSION

PHIL WARD

A RAIDING FORCES SERIES NOVEL

This book is a work of fiction. Names, characters, businesses, organizations, places, events and incidents are either a product of the author's imagination or are used fictitiously. Any resemblance to actual persons, living or dead, events or locales is entirely coincidental.

Published by Military Publishers, LLC
Distributed by Military Publishers, LLC
Austin, Texas

www.philwardauthor.com

ISBN: 979-8-9920645-0-6

Cover design by Stewart A. Williams
Map illustrations by Tom Houlihan

For ordering information or special discounts for bulk purchases, contact

MILITARY PUBLISHERS LLC
3616 Far West Blvd., Suite 117, Box 215 Austin, TX 78731

DEDICATION

The Magnificent Mission is dedicated to the men of the
1st Recondo Battalion Republic of Vietnam 1968.

COVER COLOR BY

GRACIE

RANDAL'S RULES FOR RAIDING

RULE 1: The first rule is there ain't no rules.

RULE 2: Keep it short and simple.

RULE 3: It never hurts to cheat.

RULE 4: Right man, right job.

RULE 5: Plan missions backward (know how to get home).

RULE 6: It's good to have a Plan B.

RULE 7: Expect the unexpected.

RANKS, DECORATIONS AND NICKNAMES

RANK PROTOCOL:
The first time a person is named in a chapter or after a chapter break their full rank and name is given. Addressing military personnel by their rank is a mark of respect. At all levels rank is earned and those who have it from a corporal to a four star general are proud of it.

DECORATIONS:
In the British military officers are authorized to put the initials of their decorations after their name. In the Raiding Forces Series the protocol is the first time an officer is introduced in a book the initials of his decorations are listed following his name. After that for the rest of the book they are not.

In the U.S. military officers do not have the same privilege.

NICKNAMES:
In the British military nicknames are endemic. Radio operators are called Sparks, red heads are called Ginger, tall people are called Lofty but sometimes short people are called that too etc.

In the U.S. military there are a lot of nicknames but nothing like the British.

ONGOING OPERATIONS

OPERATION BAD CALL

MI-5 Counterintelligence program — code name BAD CALL Identify Greek collaborators on the islands in the Aegean Theatre of Operations. Once an individual was targeted the SOG would dispatch a team to conduct a snatch mission to bring the traitor off the island where they lived. That not being possible, a wet team would be dispatched to assassinate them.

OPERATION GOLDEN FLEECE**

"Pinch" operations to capture Nazi encoding/decoding equipment. Lt. Cdr. Fleming's project, aka OPERATION RED INDIAN.

OPERATION LONG NECK

A secret operation to intercept diamonds being smuggled from the Congo to the Nazis and to eliminate the diamond smugglers.

Command and Control Team—code name **CARD GAME** Colonel John Randal, Major the Lady Jane Seaborn, Lt. Gen. "Geronimo" Joe McKoy, Captain Billy Jack Jaxx, Waldo Treywick, Captain Pamala Plum-Martin, Mandy Paige, Beverly Blackwell, King, Captain Roy Kidd, Captain Preston Butterfield, Master Sergeant Mack Beckwith.

OPERATION RED INDIAN

Cover name for **OPERATION GOLDEN FLEECE****

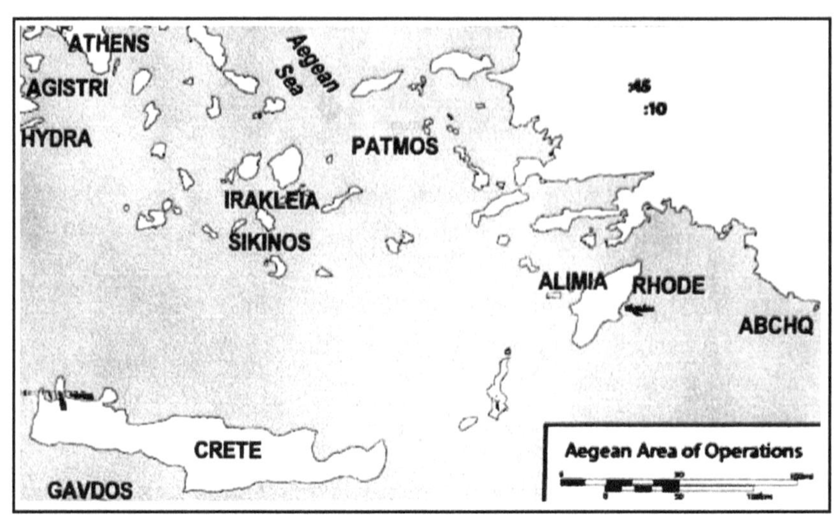
ATHENS
Aegean Sea
AGISTRI
HYDRA
PATMOS
IRAKLEIA
SIKINOS
ALIMIA
RHODE
ABCHQ
CRETE
GAVDOS
Aegean Area of Operations

1

SHOOT, MOVE AND COMMUNICATE

CAPTAIN BILLY JACK JAXX WAS IN BIG TROUBLE. WHICH was nothing new. Only this time it looked like he might not get out of it. The British submarine *Perseus* had struck a mine and gone down with him on board seven miles off Hydra, a 19-square-mile German-occupied island. Being a supernumerary along for the ride he did not know the sub's location when it sank. Vice Admiral Sir Randolph "Razor" Ransom, VC, KCB, DSO, OBE, DSC, RN had arranged a familiarization cruise for him on the sub, which was making a "Magic Carpet Service" run to Malta with an eye to utilizing submarines for future Small Raids Incorporated/Raiding Forces operations.

It was supposed to be a milk run. No one anticipated the *Perseus* being sunk. Least of all Jack Cool.

Launched from Vickers-Armstrongs Barrow yard ten years prior to the war the *Perseus* had a surface displacement of 1,475 tons—2,040 tons submerged. She had a surface speed of seventeen and a half knots and was armed with a 4-inch gun and eight 21-inch torpedo tubes—six bow, two stern. Upon commissioning, *Perseus* joined the Fourth Submarine Flotilla on the China Station. In 1940 all China Fleet submarines were recalled from the Far East to take up station in the

Aegean. The *Perseus* was assigned to the First Submarine Flotilla operating out of Alexandria. Her primary mission was ferrying vital supplies to the besieged island of Malta. These trips were dubbed the "Magic Carpet Service."

At dawn on 3 December 43 while on a Magic Carpet supply run, the *Perseus* spotted an enemy supply transport. Fed up with being a freight hauler, the skipper, Lieutenant Commander Edward "Ted" Nicolay, RN, attacked. Two torpedoes were fired, which may or may not have struck home. Before the sub could confirm results, two Kriegsmarine escorts arrived and attacked with depth charges. Lt. Cdr. Nicolay elected to crash-dive and continue the resupply mission.

Capt. Jaxx found himself seeing more naval action than he had counted on—getting depth-charged in a submarine being one of the more terrifying experiences of his military career—which had its moments.

Three days later, late in the evening of 6 December 43 while cruising off Hydra Island the *Perseus* struck a sea mine. A terrific explosion blasted the sub blowing a massive hole in her starboard bow. The submarine plunged straight to the bottom and settled on the seabed with an acute list to starboard. Capt. Jaxx, who was in the after compartment at the time, was thrown off his feet and slammed into the steel bulkhead, momentarily stunned by the violence of the detonation. When he regained his senses the boat was completely blacked out. At that point Capt. Jaxx was wondering what could have possibly motivated him to explore the joys of submarining. Sounded like a good idea.

After searching around blindly he found the rack of emergency flashlights—aka torches in Royal Navy parlance. He had been leaning against them before the explosion. Switching one on he discovered five badly shaken stokers. With a great deal of effort the sailors were able to secure the watertight door. Unfortunately, when it was shut the six were confined in a small space and oxygen was rapidly running out. To make matters worse, fumes were coming from the drums of oil and enamel paint stowed in the compartment. The containers had burst from the explosion and toxic gasses were rapidly poisoning what little breathable air there was.

Prior to sailing Capt. Jaxx had the benefit of about ten minutes' training on how to escape from a sunken submarine utilizing gear known as the Davis Submerged Escape Apparatus (DSEA) for just such a contingency. It was designed to enable the wearer to breathe under water long enough to abandon ship and swim to the surface. It included a metal clamp to clip on your nose which did not do much to inspire confidence in anyone thinking about having to actually employ the DSEA.

The reason instruction on the apparatus was so brief was because the Royal Navy considered the odds against escaping from a sunken submarine extremely high for experienced sailors. No army tourist was likely to be able to accomplish it.

One of the sailors managed to undo the escape hatch, pull down the canvas accordion-like trunking, and secure it to the deck. Special flood valves were fitted in submarines beneath the hatches on canvas trunks that could be pulled down. The idea was to flood the compartment and allow the water to float the sailors up the shaft to make their escape. Clearly, the DSEA was an escape method designed by a lunatic.

The acute angle of the submarine caused the starboard side hull to become the deck—and the rapidly-foul smelling air made escape efforts deemed virtually impossible by the Royal Navy even more difficult than normal. Having to work by flashlight while breathing poisonous fumes did not make things any easier. Three of the sailors were dead by the time the trunking was secured and another was on the verge of lapsing into unconsciousness.

The trick now was to get out of the compartment and up the tubelike trunking. From his briefing, Capt. Jaxx knew the only way to escape was by flooding the room which sounded like a really bad idea now that he was actually faced with having to do it. And then he would have to float himself up the shaft until he popped out at the top. Well, that was insane. He and the remaining stoker searched around under the oily water until they finally located the flood valve—only to discover it was bent and could not be turned.

The sailor managed to croak, "Underwater gun."

Then he passed out.

The underwater gun was a small vertical torpedo tube used by submariners in distress to send flares to the surface. Now all alone Capt.

Jaxx found the wheel on the underwater gun and cranked it open. The oily slime in the compartment began to rise as seawater rushed in. So far so good, but on his first attempt to pass through the narrow trunking to the surface he was forced back down the canvas tube by a powerful blast of dirty air.

Desperate, bordering on blacking out himself, Capt. Jaxx tried again. This time he was successful managing to shoot up the tube without fouling the submarine's jumping wire—he had no idea what that was but it had been mentioned in his escape briefing as something to be concerned about. Next problem: if the *Perseus* standard operating procedure (SOP) had been followed all the hatches except for the conning tower's would not only have been bolted internally but also secured by a steel bar externally to prevent the possibility of one of them popping open when the sub was being depth-charged.

Escaping out the hatch in the after compartment was going to be impossible and there was no way to reach the conning tower. It was probably a good thing Jack Cool was not read in on all the *Perseus'* SOPs, which covered securing the hatches on war patrols so he did not know about the crossbar as he shot up the funnel.

He was lucky. Tonight some sailor failed to do his job. The aft hatch was left unsecured.

Capt. Jaxx made it out of the *Perseus* but he was a long way below the surface. That was a problem. He had learned to swim in a West Texas stock tank which meant not much water over his head. Fortunately, Rikke Runborg had been conducting a daily two-mile swim for the female members of Raiding Forces. The ice-blond Norwegian bombshell loved to tease him so she had issued an open challenge to join the women with the promise, "if you catch me you can have me." His swimming skills had definitely improved over the last year, though he never caught Rocky.

By the time Capt. Jaxx popped to the surface, having been briefed during the abandon ship orientation to go slow to avoid getting the bends—whatever those were—his lungs felt like they were about to explode. He had forgotten to put the clip on his nose.

The sea was calm. He was all alone surrounded by bubbles rising and bursting silently from the wrecked sub below, and that was spooky. The *Perseus* had gone down with all hands.

Capt. Jaxx was the sole survivor of the fifty-three-man crew.

In the far distance was a dark smudge that might be a landmass. With no other option he could think of Capt. Jaxx started swimming in that direction. Unknown to him he was headed toward Hydra Island, a place some claimed was named after the monster in Greek legend described as a giant water snake with at least nine heads—one of which was immortal.

Jack Cool's problems were not over.

COLONEL JOHN RANDAL AND MAJOR THE LADY JANE Seaborn, LG, OBE, RM, were outside of Advanced Base Castelrozzo Headquarters (ABCHQ) making an inspection tour of the class underway for the officers of the Small Boat Squadron (SBS) preparatory to the commencement of a selection course. Lieutenant Chase Starrett was accompanying them. As was Happy, Lady Jane's German shepherd.

The SBS officers were broken down into small groups so they could receive individual attention during the period of instruction. Each group was being taught by a Raiding Forces operator—officers and NCOs selected by Captain Roy Kidd for their experience and communications skills.

Today's subject was "Warning & Operations Orders" . . . there would be a practical test later.

Lieutenant General "Geronimo" Joe McKoy was standing off to the side smoking one of Waldo Treywick's custom-rolled cigars observing the first class they came to. It was being taught by Sergeant Jerry Schmidt, formerly of the 10th Ranger Battalion, now a serving member of Captain Billy Jack Jaxx's Small Operations Group (SOG). Training preparatory to the reorganization of a military unit was something Lt. Gen. McKoy had a wealth of experience with and was personally and professionally interested in. As was Lieutenant Colonel Sir Terry "Zorro" Stone, KBE, DSO, MC, who arrived a few minutes later.

Following the noon meal the SBS officers would reassemble and be issued a Warning Order for a training raid each of them would lead later that night. At the end of class the officers would marry up with teams composed of troops from the 1st Battalion King's Own Royal Regiment (KO)—former Habbaniya Strike Force men, who were waiting for transport to No. 4 Middle East Training School (METS) at Kabrit for their indoctrination into the gentle art of military parachuting—and Office of Strategic Services (OSS), Operational Group (OG), Detachment 3 (Det 3) Frogmen recently returned from their first missions with Raiding Forces. At that time each of the SBS officer candidates would issue their individual Warning Orders to their teams.

The shortage of amphibious air transport available—due to the Hudson being grounded for repairs because of damage incurred on an unauthorized low-level bombing run against a Luftwaffe landing ground on Rhodes—meant the KO's student paratroopers would have to be flown out of Advanced Base Castelrozzo (ABC) in shuttles. Which worked out perfectly for Capt. Kidd's purposes. He could use them to compose teams so the SBS officers could lead the simulated raids later tonight—there not being enough Frogs to go around.

Following their Warning Order each of the Raid Team Leaders would then issue his own Warning Order to his team. Then the SBS officers would go through the Troop Leading Procedures as the men prepared for the mission. After an appropriate period of time, again as if it were an actual operation, the process would start once more with Operations Orders being issued. Finally, after nightfall the SBS officers would load out their teams on caiques and lead them against various targets located around the perimeter of Castelrozzo Island simulating raiding from the sea.

Everything the Raid Team Leaders did from the moment they linked up with their team would be graded. It was PASS/FAIL. No in-between. Each SBS officer hopeful would be required to lead six training raids. To successfully complete Capt. Kidd's selection course, a requirement for retention in Raiding Forces, an officer had to pass 50 percent of his raids.

Sounded easy enough, 70 percent being the typically accepted standard for a passing grade, but that was misleading. The anticipated class failure rate was 60 percent or higher for commissioned officer students—all of whom were special operations veterans. To make the cut a candidate not only had to be an outstanding leader of men and a master of small unit tactics, he also had to be an individual who would be able to assimilate into the Raiding Forces culture. A colorful personality was not a drawback. Being a legend in their own mind or a know-it-all was a deal killer.

Even though Col. Randal had a shortage of troop commanders he was only taking those who demonstrated the dual ability of being team players while also capable of leading independent operations against long range targets. A rare blend of contrasting leadership styles. He wanted officers and NCOs who would be the best fit for Raiding Forces no matter how much time they had spent in combat or how many operations they had gone on in the past.

Pass 50 percent of your raids or be returned to your unit (RTU). Oh, and you have to gain our respect while you are at it.

What appeared to be a simple no-frills selection program was in fact a sophisticated multilayered exercise designed to accomplish much more than merely weeding out those judged not qualified to serve in Raiding Forces. Col. Randal was intent on building a cohesive Special Operations unit—not simply amassing the remnants of fabled raiding/reconnaissance outfits who had seen better days. He was only going to take officers, NCOs and men who would be a good fit.

SBS officers were undergoing selection on ABC under Capt. Kidd while SBS NCOs were holding their selection trials at the base outside Atlit, Palestine, under Master Sergeant Mack Beckwith. As the troops for the teams the NCO's would lead the Sergeant Major was utilizing paratroopers from the 11th Parachute Battalion (PB) who had volunteered for Raiding Forces. Some from A Company were veterans who had dropped on Kos.

Once selection was complete the officers and men of the SBS, 1st Bn KO, OSS OG Det 3 and 11th PB would compose one of the best intakes of new personnel Raiding Forces had ever enjoyed. Just one problem: there were not enough of them.

Col. Randal's idea was for the instructors to evaluate the officer and NCO candidates and in the process begin developing relationships with them that would pay dividends later. Additionally, other interested parties in Raiding Forces, of which there were many of all grades, were welcome to come observe selection to judge for themselves how well the aspirants were performing. Simultaneously, the SBS officers and NCOs would be assessing the KO and 11th PB volunteers to identify men they might want to recruit when they organized their own teams.

As would the Raiding Forces personnel who were observing.

Meanwhile, the KO and 11th PB officers and men were getting an introduction to operating in small teams instead of the conventional line infantry table of organization and equipment (TO&E) of companies, platoons and squads. And the new volunteer troops would be forming their own opinions, deciding which officers they wanted to indent to serve under.

Col. Randal had long maintained a policy of "putting round pegs into round holes" and this was one way to do it . . . letting the new Raiding Forces personnel fill out a "dream sheet" for where they would like to be assigned within the unit. It was important for all those involved with selection to perform at their highest level—instructors, officer/NCO candidates, and new individual volunteers. Everyone was being evaluated on some level by someone. Selection is the equivalent of inside pool on how to build a Special Operations unit. It was a complex process. Only a true professional would recognize the method to the madness.

But not all of them.

Lt. Gen. McKoy said, "I understand Major Patterson's conductin' his own shakedown of the SBS squadron he just assumed command of."

Col. Randal said, "His officers and NCOs are going through Captain Kidd and Sergeant Major Beckwith's selection courses. Major Patterson's evaluating the troops."

Lt. Gen. McKoy said, "Word is Ian's RTU'n people left and right."

Col. Randal said, "That's what I hear."

CAPTAIN BILLY JACK JAXX SWAM FOR SIX HOURS according to his new Rolex—good thing it was as waterproof as advertised. Before striking out he had held his flashlight against the watch's face to charge the luminous green digits marking the numerals and the hands to make them glow brighter for better visibility in the dark, then dropped the light in the submarine to lose the extra weight. It was going to be hard enough going in his battle dress and canvas-topped raiding boots. He alternated freestyle for short distances with breast strokes, finally settling on the breaststroke. The DSEA gear strapped to his chest now served as an awkward life vest that kept him afloat but hindered serious swimming efforts. He had been asked to leave his weapons behind at the Alexandria office of Small Raids Inc., so at least he did not have them weighing him down.

However, he had not entirely complied with the Royal Navy's request. There was a .32 Savage 1907 in the right billows pocket of his M-1942 jump jacket, which was a good place to conceal a pocket pistol endorsed by no less lights than Bat Masterson and Buffalo Bill Cody. The handgun had been his backup piece since he served warrants with his grandfather, the Sheriff, back home in Texas when he was in high school. In his left side pocket were two spare double stack ten round magazines—a radical development in firearms engineering for the time period the Savage 1907 was manufactured.

A turn-of-the-century photo of a chubby slightly over-the-hill Bat Masterson wearing a snazzy straw boater and a bow tie appeared in advertisements in some of the old gun magazines lying around the sheriff's office quoting him saying, "Ten shots quick, as fast as you can press—*not pull*—the trigger." The former Army scout, lawman, buffalo hunter, gunfighter and New York newspaper opinion piece writer knew quite a bit about pistol shooting technique. Especially the importance of trigger control. Something he felt strongly enough about to be compelled to emphasize it to less experienced gun purchasers in the advertisement for the Savage 1907.

The little pocket blaster and the M2 Paratrooper switchblade knife zipped in the collar of his jump jacket were the only weapons Capt. Jaxx carried. They felt like an anchor but he might as well have drowned rather than swim to a likely enemy-occupied island unarmed. At least that was his estimate of a really bad situation. Still, the idea of invading a Nazi-held island armed with a small-caliber pocket pistol and a switchblade knife had nothing to recommend it, no matter what Bat Masterson said.

Finally Capt. Jaxx washed up on land totally spent, nearly dead. Not having the strength to walk, he crawled ashore and discovered a small cave just off the beach, the kind commonly found along the coastline on Aegean islands. He made it inside and passed out from exhaustion. At sunrise two Greek fishermen who came walking along the shore spotted signs that looked like a giant snake had slithered across the tiny beach. They cautiously followed the track with visions of the nine-headed Hydra running through their heads—though the local men were pretty sure the monster was only a fable.

Capt. Jaxx came to with a start when the two peered inside the cave but did not show the Savage .32. He did not speak much Greek, only a few words Alex “Cat” Gataki had taught him, and those had no value in a situation like this. The Greeks did not speak English. The fishermen thought he was a Nazi counterintelligence agent clandestinely infiltrated to root out islanders who might be resisting the German occupation of Hydra. The Abwehr was known to do such things. Much jabbering and waving of hands ensued. It took a while for things to calm down.

Capt. Jaxx kept saying, “American, American . . .”

The fishermen kept insisting, “Germanous, Germanous . . .”

Which sounded a lot like “German” to Capt. Jaxx.

Finally Capt. Jaxx showed them the label on the underside of his M-42 Jump Jacket. The fishermen relaxed when they saw “Jacket, Parachute Jumpers,” written in English. The Greeks indicated they wanted him to remain in the cave while they went for help.

Capt. Jaxx did his best to communicate the importance of not informing any Germans on the island of his presence.

The fishermen emphatically shook their heads they would not do so. But, of course, they did.

EVEN WITH MOST OF THE RAIDING FORCES TROOPS AWAY there was a lot going on at ABC. The unit was in the process of a major reorganization necessitated in large part by the heavy losses incurred by the 575th Ranger Task Force on the Benevento drop following the invasion of Italy. At least a third of the paratroopers who made the jump were still MIA. An estimated 10 percent had been killed with an additional 30 percent wounded. Rangers were continuing to straggle back to friendly lines and being returned for duty. So the exact extent of the losses was still to be determined.

Major the Lady Jane Seaborn, Captain Pamala Plum-Martin, DSO, OBE, DFC, RM, Mandy Paige, OBE, and Beverly Blackwell were in the initial stage of organizing an MI-5/OSS X-2 counterintelligence operation—OSS called counterintelligence "Counter Espionage." The idea was to identify Greek collaborators on the islands in the Aegean Theatre of Operations (ATO). Once an individual was targeted Captain Billy Jack Jaxx would appear unannounced out of the dark of night with a SOG team to conduct a snatch mission to spirit the traitor off the island. That not being possible, a wet team would be dispatched to assassinate them. Major Zargo was providing a Greek Sacred Squadron (GSS) liaison officer to facilitate GSS cooperation on the project. Capt. Jaxx had suggested BAD CALL as a code name for the counterintelligence operation. Meaning the traitors had made the wrong choice when they decided to side with the enemy. It was as good an identifier as any since it had no connection whatsoever with anything else Raiding Forces was doing. And it would likely make no sense to the Abwehr, the German intelligence forces.

Getting BAD.CALL up and running was proving challenging. Other demands had been placed on the men and women organizing it as well as on the troops involved in active and supporting roles. For example, Beverly had been called away to fly in the new Hudson obtained for Raiding Forces by her father, Major General Sam Houston Blackwell—friends called him "Bronc." He had arranged for the replacement because he was responsible for getting the old Hudson shot up on an unauthorized bombing mission joyride.

Capt. Plum-Martin and King were on a LONG NECK diamond-buying trip to Morocco. And Capt. Jaxx—so far as anyone at ABC knew—was observing submarine operations incidental to determining how to employ them for clandestinely inserting Raiding Forces teams on targeted islands in their Area of Operations (AO).

Lieutenant General "Geronimo" Joe McKoy and Waldo Treywick were finalizing details for a trip to the Belgian Congo to inspect Commander General Frank Polanski's LONG NECK diamond interdiction program along the Congo River. While they were away, Capt. Kidd—in addition to supervising the Raiding Forces Officers Candidate School (OCS)—was to assume their duties chasing down rogue smugglers attempting to circumvent The Three's monopoly of the illicit diamond trade. From time to time on request he would work in conjunction with Captain Preston Butterfield III to interdict smugglers out of Cairo who were attempting to run camel caravans to Turkey. Except he had been pulled off that project temporarily to organize the Small Boat Squadron officer/NCO's selection courses.

Major Butch "Headhunter" Hoolihan, DSO, MC, MM, RM, was in the process of retraining Commander Ian Fleming's Naval Intelligence Division 30 Assault Unit aka Red Indians, which was complicated by the Royal Marines in the unit having to go through selection. Not many of them were going to be retained in Raiding Forces. That was fine with the Headhunter as the losses would be made up by Marines brought in from the troop he had led for over a year in coastal pinprick raids against the Via Balboa.

Lady Jane was in the TOC reading a message from her godfather, Lieutenant Colonel John Henry Bevan, the "Chief of Deception" in command of the London Controlling Station (LCS)—even the name of his organization was classified MOST SECRET/TOP SECRET/ NEED TO KNOW. Only a handful of people possessed that need. He was requesting a meeting with Colonel John Randal and Lady Jane in London before the end of the year—only weeks away.

Mandy Paige, already the most overtaxed counterintelligence officer in Raiding Forces, would have to be tapped to stand in temporarily for Lady Jane to continue organizing BAD CALL in her absence. Capt. Jaxx was delegated to support her for SOG quick reaction missions, both snatch and wet work—when he returned from his excursion onboard the *Perseus*.

James "Baldie" Taylor was in Vice Admiral Sir Randolph "Razor" Ransom's office briefing the Razor and Col. Randal on new developments in store for future operations in the ATO. The developments would have far-reaching ramifications for Small Raids Inc. and Raiding Forces. This meeting was classified.

Jim said, "As a result of bruised feelings about the clumsy way Middle East Headquarters, Plans Division has handled Special Operations. And SOE being notorious for poor security, infighting and petty conflicts with other agencies here in the ATO—a decision has been made to reappraise raiding responsibilities. In short both organizations are having their wings clipped."

Col. Randal glanced at VAdm. Ransom. The expression on the Razor's face approximated what he imagined a smiling man-eating crocodile would look like. There was no love lost between the commander of Small Raids Inc. and the Plans Division or SOE.

Jim said, "Effective immediately, all small unit naval assets will fall under your control, Admiral. For example, Commander Seligman's Levant Schooner Flotilla will now report to Small Raids Inc. as will the 10th MTB Flotilla and other similar Royal Navy and independent irregular outfits that have sprung up like mushrooms in the ATO. Deploy them as you see fit. In the future SOE has to come to you with a request for transportation to insert or extract agents by sea and/or to transport their clandestine parties on whatever nefarious mission they have dreamed up. In addition Plans Division has to request naval support from Small Raids Inc. for any and all projects they have in mind."

VAdm. Ransom said, "Outstanding! We shall tighten up irregular small boat operations in short order. You responsible for this, General?"

Dodging the question, Jim said, "All raiding activity in the Aegean shall now fall under Raiding Forces. In previous conversations, Colonel, you indicated a desire for SOE to have responsibility for the larger islands and for Raiding Forces to handle raids against the small islands—quick ins and outs hitting and running. That has been approved with one caveat. Now you shall also be able to conduct coastal raids against targets on the larger islands if you so choose—you sign off on all raiding operations to include those developed by SOE or Plans Division to ensure coordination."

Col. Randal said, “Works for me.”

Jim said, “Under this new division of responsibilities while all raids must clear through Raiding Forces, SOE shall not be required to disclose the classified nature of its missions. You do not have the authority to cancel SOE operations, Colonel. Only to coordinate them so there are no blue-on-blue conflicts due to the left hand not knowing what the right hand is doing . . . as has occurred in the past with unfortunate results.

“That stipulated, this development gives you enormous command and control over small-scale raiding in the ATO. Now you will be able to implement your Constant Pressure Concept free from outside interference. I have been designated to arbitrate in the event a protest is lodged by either Plans Division or SOE—we all know that is inevitable.”

Col. Randal said, “Good.”

Jim said, “Brigadier Turnbull shall continue to be listed as Chief of Raiding Operations Middle East Command—on paper. As you are aware, in the British Army brigadier is a *title,* not a *rank*. It is intended to make the man holding it senior to other colonels . . . our brigadiers are not generals. The U.S. Army does not have the same policy. You Americans do not recognize British brigadiers as automatically outranking your full colonels.

“Turnbull is not senior to you and he is not in your chain of command. His post is merely symbolic. In fact, you do not even have to allow him on ABC if you so choose. Acting on my security recommendation, Admiral Ransom has formally declared Castelrozzo a restricted island. A black site, if you will. The less outsiders know about what we do here the better.

“For the record, the Admiral is in *your* chain of command.”

Col. Randal said, “I hear you loud and clear, General.”

VAdm. Ransom said, “Quite an interesting development.”

Jim said, “Thought you would approve.”

VAdm. Ransom said, “Small-scale raiding operations in the ATO, which thus far have been planned by Middle East Headquarters with about as much strategic forethought as throwing darts at a map board blindfolded, are about to experience a Force 5 shakeup. Standby ready, General. Your mediation skills are going to be put to the acid test.”

Jim said, “There are not going to be any mediations.”

As the men left the Razor's office, the Admiral's new WREN—Lieutenant Coco Lovejoy, recently transferred to Lady Jane's Royal Marines to be Beverly's assistant on OPERATION BAD CALL—said, "Sir, Terry requests the pleasure of your presence in his office at your earliest convenience, Colonel."

THE TWO GRAY-HAIRED GREEK FISHERMEN CAME walking back along the beach. Only this time they had a German feldwebel (sergeant) and two privates with them. The three Nazis were wearing the cap insignia of the 1st Mountain Division—a pretty Edelweiss flower, indicating they were Gebirgsjägers, or "mountain light infantry hunters," or just "mountain hunters." Typically they were called 'Jägers.'

The 1st Mountain Division was an elite formation composed of some of the most battle-hardened troops in the Wehrmacht. The 1st Mountain Division had fought in Poland, France, Yugoslavia, Russia and the Balkans prior to their transfer to the ATO. Elements of the division executed 6,000 Italians of the 33rd "Acqui" Division *after* they surrendered on Cephalonia Island. The 1st Mountain Division Gebirgsjägers were even more cold-blooded killers than the 999th Light Infantry Division (Criminals). The difference being the 999th were poorly trained convicts let out of prison to do their dirty work against the civilian population in occupied areas to create terror while the 1st Mountain Division consisted of elite, battle-tested, alpine troops—vicious stormtrooper Jaggers who had fought it out in open battle against the Third Reich's toughest adversaries. And had no qualms about shooting prisoners *en masse*.

The feldwebel was armed with a 9mm Walther P-38. One of the privates was carrying a Waffenfabrik Steyr 9mm MP-34 submachine gun and the other a self-loading Russian Tokarev 7.62 SVT-38 rifle. When the party approached the cave, Captain Billy Jack Jaxx stepped out from behind a boulder with his .32 Savage 1907 pocket pistol in hand—a weapon that looked a lot like a Buck Rogers ray gun—and shot the three mountain hunters in the back of the head. One round each. *BANG! BANG! BANG!* The gunshots sounded sharp in the early morning air. The Germans dropped in their tracks as if pole axed. Bat Masterson would have been impressed.

The two Greeks whirled around in alarm—he shot them too.

Capt. Jaxx walked over and searched the dead Nazis, not finding much of interest other than their weapons and packets of matches for their Sturm cigarettes. The brand name meant “Military Assault,” or so he had been told, not speaking a word of German. He buckled the 9mm Walther P38’s pistol belt around his waist, slung the 7.62 SVT-38 over his shoulder and appropriated the 6x30 Zeiss binoculars the feldwebel had been carrying in a brown leather case. Then he draped the bandoliers containing the ammunition for the weapons across his chest like a Mexican bandito and put the Model 24 Stielhandgranates, aka “potato masher,” stick grenades he took from the privates’ jackboots into the reinforced billow pockets of his US M-42 paratrooper jump jacket.

Each of the ski troops carried a unit-specific, special-issue stag horn-handled knife peculiar to the 1st Mountain Division that looked more like a steak knife from an exclusive restaurant than a military fighting knife. The division’s name was stamped on the five-inch drop-point blade. Had to be a story there but Capt. Jaxx had no idea what it was. The knives were a strange choice for a fighting weapon if that’s what they were. But he added one to his belt and the other two to a bandolier across his chest. The knives were also unusual in that the bumpy stag horn handles were left roughed out—not sanded down smooth as was normally the practice. It looked like gripping them hard could hurt.

Capt. Jaxx picked up the 9mm MP-34 submachine gun, racked the charging handle far enough to make sure a round was chambered, and decided it best to move out to other environs with all due speed. Well briefed on enemy forces in the ATO, he knew the typical German Army squad was composed of nine men led by a corporal. In a light infantry Jäger formation like the 1st Mountain, a squad might consist of one or two fewer men. However, the presence of the feldwebel indicated the possibility of more than a single squad on the island.

What it all meant was Capt. Jaxx was alone on an island he did not know the name of or how large it might be—or even if it was, in fact, an island. For all he knew, he could have washed up on mainland Greece. Initial indications were he was facing at minimum eight to ten highly trained, battle-hardened killers not shy about getting blood on their hands. Likely there were more than that. The Gebirgsjägers would be coming for him sooner rather than later and they were not going to be happy about losing three of their own—shot in the back of the head.

Maybe he should have thought through that part a little better.

COLONEL JOHN RANDAL WALKED INTO LIEUTENANT Colonel Sir Terry "Zorro" Stone's office. "Looking for me?"

Lt. Col. Stone said, "There is something I have been needing to talk to you about for a while now but put off. Not looking forward to the discussion."

Col. Randal clicked on. This was not the typical start of a conversation with his best friend. It did not bode well.

"OK, now would be the time . . ."

"As you are aware my father has been an ardent supporter of all things Raiding Forces. Took a while to win him over when we first brought in the Lancelot Lancers Yeomanry. He envisioned them being the reconnaissance unit for some armored division or independent armored cavalry brigade riding into battle with guidons flying. But once the Duke saw the way the regiment was being utilized, he came on board in a big way.

"Now that has taken a new twist."

Col. Randal said, "I see."

Lt. Col. Stone said, "Father has it in his head that when the second front opens up, the Lancelot Lancers shall be in the van of the first wave storming the beaches of Fortress Europe. He has asked for them to be brought home to prepare for the invasion."

Col. Randal said, "Your family regiment has been dramatically reduced over the last two years. It was the smallest on the British Army's Yeomanry list to start with—only one troop. The Duke's home county's inability to supply replacements hasn't been able to keep up with attrition."

Lt. Col. Stone said, "That is a fact, old stick."

Col. Randal said, "The Lounge Lizards are one of—if not *the*—most capable squadron we have. Your cousin Mongo's a first-class officer. Losing them . . . that's not good, Terry."

Lt. Col. Stone said, "Once the Duke fixates on a thing there is no dissuading him. The family regiment has to lead the Great Crusade against the evil Nazi Empire—land ashore and drive on Berlin. Stone family honor is at stake."

Col. Randal said, "Yeah, I can understand how your father might feel that way. So, when does the Duke want his people home?"

"By the first of the year . . . he expects me to resume command."

"Lovely."

Lt. Col. Stone said, "Have you heard the latest news flash from Hollywood?"

Col. Randal said, "Negative."

Lt. Col. Stone said, "They have changed the name of the movie based on our snatch mission of the Italian admiral. Instead of *Jump on Bela,* now it is to be called *Death from Above.*"

Col. Randal said, "Seeing as how no one on either side was killed . . ."

Lt. Col. Stone, "If I ever come across Captain Dennis Wheatly he may become the only casualty for writing that book."

Col. Randal said, "Be my guest."

CAPTAIN BILLY JACK JAXX ASSESSED HIS PERSONAL situation. He did not know where he was—but no matter the location it was deep behind enemy lines. He had no food. No water. Did not speak the local language. He was up against a detachment of 1st Mountain Division Jägers of unknown strength. The unit consisted of veterans of half a dozen theatres of operations. It had been accused of committing atrocities against civilians, ethnic cleansing and/or other war crimes in the past. He had just shot three of the division's troops in the back of the head, so there went the Geneva Convention.

On the bright side, he had a lot of weapons now.

What to do? He knew the direction the Greek fisherman and 1st Mountain Division Jägers were traveling as they approached the cave—useful intel. The terrain, as far as he could ascertain, was rugged—hilly, rocky and sparsely covered with pine and what appeared to be maple trees. There was not enough vegetation to offer much in the way of cover and concealment except in the small valleys. In the distance could be seen a respectable-sized volcanic mountain that appeared to be 2,000 feet tall or so. It was the type of terrain feature best to keep away from. Particularly considering the opposition consisted of mountain climbers. Prominent topography in general is always something to avoid when on the run behind enemy lines. You do not want to be anyplace the eye is naturally drawn to.

And there was something else worth consideration. The three Germans, while heavily armed, had not been wearing any load-bearing gear or carrying canteens. To Capt. Jaxx it indicated the three ski troopers were not far from their base camp and had not intended to be away for long. Bring in an unarmed man—how hard could that be?

The island—Capt. Jaxx was "presuming," not "assuming" it was an island as "assume" was a banned word in Raiding Forces—did not look like it could sustain flocks of sheep. At least not this part. Which meant there were not likely to be sheepherders, who were usually teenage boys. Inquisitive youngsters roaming the countryside can pose a real problem for an evader. Being Greek they might help or maybe not. Isolated farms, while probably rare in this region, could be a source of food, water, and information and possibly provide a place to hide out. Never-the-less they would need to be approached cautiously.

The typical Greek village in the ATO was a replay of Castelrozzo—an amphitheater of houses cascading down the slope to the local harbor. On an island this barren, sponge fishing had to be the main source of income. Possibly he could steal a caique. Only he would not have any idea what direction to sail, even though he was wearing his wrist compass. The problem with navigation being if you do not know where you are to start with there is no way to know which way to go . . . and Capt. Jaxx did not have a clue where he was.

Besides, navigation by the stars at sea requires specialized training and certain instruments. Capt. Jaxx had neither. He had taken a class in Officers Candidate School (OCS) at Ft. Benning that taught the basics of celestial navigating—though he had not understood a word of it. When the students went outside and looked up at the night sky he could never visualize the various celestial formations the instructor was pointing out other than the Big Dipper. There were a lot of twinklers up there.

Land navigation, day or night, map, compass, azimuth, pace count—he was the best in the business. Problem was he did not have a map.

Other than the newly acquired weapons, what he had going for him was that this place had in all likelihood been garrisoned by the Italians prior to the Germans, so for the locals it had probably been a long occupation by brutal troops. Capt. Jaxx had never been on a single island where the locals enjoyed the occupation experience. Most Greeks, meaning 99.9 percent, hated their Italian and German occupiers enough to kill them if they got the chance and thought they could get away with it. The locals might provide him aid. There were no guarantees, but the odds were in his favor.

At least that's the way it should be. The fishermen leading the 1st Mountain Division Jägers to his hiding place was not the norm. Why had they done it? Did the Germans know about the *Perseus* being sunk? Was there a reward for survivors? He was going to need to proceed with caution when it came to making contact with the locals until he had a better grip on the situation.

And there was something else Capt. Jaxx had to consider. No one was coming to rescue him because they did not know he was alive. His alma mater, the Infantry School at Ft. Benning aka "The Benning School for Boys," taught OCS candidates to "shoot, move and communicate." He was good for two out of the three. There was likely not going to be much communicating in his immediate future. So, backtracking in the direction the Germans had been traveling in order to discover where their Command Post was located seemed like the next right thing to do.

Recon the opposing forces to determine what he was up against. He was good at that.

This was starting to get interesting.

MAJOR THE LADY JANE SEABORN SAW COLONEL JOHN Randal coming out of Lieutenant Colonel Sir Terry "Zorro" Stone's office from across the room in the Tactical Operations Center (TOC). One glance and she knew something was not right. The two were so close they were able to finish each other's sentences . . . they frequently read the other's thoughts.

Lady Jane came over immediately, "What is wrong, John?"

They two stepped off to the side of the TOC. "The Duke is recalling the Lancelot Lancers back to England for the invasion—wants Terry to resume command."

Lady Jane said, "How terrible. You two have served together for ages. Even before I came along and ruined your skirt chasing."

Col. Randal said, "Never saw this coming . . . don't know how to replace him."

"I have confidence in you, babe."

"Terry's a one-off."

Lady Jane said, "I realize this is not the best timing but Uncle Johnny has requested a meeting with you in London before the end of the year. Do not ask me why. What should I reply?"

With the loss of the Lancelot Lancers looming and a reorganization of new units being assigned or attached to Raiding Forces in progress, this was the last thing Col. Randal wanted to hear.

However, prior to his recent departure from Castelrozzo, Brigadier General William "Wild Bill" Donovan had instructed him to do everything in his power to insinuate himself into Lieutenant Colonel John Henry Bevan's London Controlling Station, whatever that was. Even the name was classified MOST SECRET/TOP SECRET NEED TO KNOW and Wild Bill claimed *no one* possessed the need.

Information Col. Randal had not shared with Lady Jane—Lt. Col. Bevan being her godfather.

Col. Randal lied. “A trip to London. Sounds good. Let’s do it.”

2

NO DELEGATING THIS ONE

CAPTAIN BILLY JACK JAXX WAS IN DEFILADE ABOUT A half mile off the coastline he was paralleling, barbecuing a nice cut of cabrito—as they call roasted goat in Texas—over a small fire on the sharpened end of a stick. He had been able to start the fire with the matches he took off the 1st Mountain Division troops sent to capture him earlier that morning. Those little steak-looking knives the light infantry mountain hunters carried had come in handy too. As expected, no flocks of sheep or herders had been encountered, but there were a few free-ranging goats roaming around. The Steyer MP-34 submachine gun had a selector switch. On semiauto the sound of a single 9mm round was negligible. The goat he stalked dropped like a rock. It was doubtful anyone hearing the gunshot from any distance would recognize it for what it was.

Not that it mattered to Capt. Jaxx. He had not eaten for nearly two days and based on his experience, the quality of navy food—at least on submarines—was highly overrated. If any Nazis showed up they were going to have to fight him for his chow.

He liked roast cabrito.

So, what was the plan? Jack Cool did not have one, not really. In simple military terms, his idea was to locate the 1st Mountain Division's base camp, conduct a detailed reconnaissance of their dispositions and develop the situation.

He could do that.

COLONEL JOHN RANDAL WALKED OVER TO CAPTAIN Stephanie Fawcett-Tatum's desk. The tall glamorous brunette with the long eyelashes and perpetually sleepy eyes could have had an office of her own. However, she preferred to be in the center of the TOC where people would have easy access to her. And from where she could keep an eye on everything going on.

"My office."

Capt. Fawcett-Tatum, RM, looked up. "Yes, John."

Col. Randal shut the door behind them. "Have a seat."

Instead of sitting behind his desk he took the chair next to her. Capt. Fawcett-Tatum had helped nurse him back to health after he was shot during the Gunfight at the Blue Duck. She was a captivating woman—tall, exquisitely poised, full-on eye contact never blinking, deadly beautiful, not unaware of the effect she had on men and not hesitant to exploit that effect. The two had developed a relationship over the years based on trust and respect.

Col. Randal said, "Terry's taking the Lancelot Lancers back to England to prepare for the invasion. You two made the perfect team from day one at Oasis X. So, I'm asking, Stephanie . . . who do you recommend as his replacement?"

"I presume you require a British officer?"

"Considering the state of Allied military politics in the Aegean Theatre of Operations that would be a definite Roger."

"Percy shall do nicely. I can work with him. The two of us have known each other since childhood. We merely need to be careful not to allow him access to explosives."

Col. Randal said, "Would you be interested in the job for yourself?"

Capt. Fawcett-Tatum said, "How thoughtful of you to ask, John. You have the often-stated policy of putting round pegs into round holes. Percy is the best candidate to replace Sir Terry—I love what I do."

Col. Randal said, "You've been taking care of me a long time, Stephanie. Now would not be the time to slack off."

Capt. Fawcett-Tatum said, "Not to worry. Jane and I are there for you."

"Yes you are."

VICE ADMIRAL SIR RANDOLPH "RAZOR" RANSOM WAS IN his office. He was reflecting on the implications of light surface units in the Aegean Theatre—what the Royal Navy characterized as "coastal forces"—falling under his command while reviewing a list of the units now assigned to Small Raids Inc. Being the "Admiral Commanding Coastal Forces Aegean" was a dead-end assignment. The Royal Navy was a big-ship navy. Battleships, aircraft carriers—with submarines the exception—were the commands that carried status. Small surface units, even though some were not so small, were derisively referred to as the "dust of the sea." Still, considering VAdm. Ransom had been brought out of retirement to serve, he was extraordinarily pleased with the development.

The enormity of the task at hand was not lost on him. Creating a unified command out of so many disparate units spread over the entire Aegean operating from dozens of bases was going to be a challenge. As Lieutenant General "Geronimo" Joe McKoy was known to say, "like tryin' to herd cats." VAdm Ransom considered it an appropriate analogy. The transition would require serious prior planning and a firm hand at the helm. And while there were a lot of moving parts, none were very large, which at first blush might seem like a good thing—it probably was not. What was immediately clear: he was going to require a bigger Small Raids Inc. staff footprint on ABC. The logistics of sustaining the SRI small unit fleet were going to increase exponentially. Looking forward there was a lot of work to be done and when it was complete, Small Raids Incorporated (SRI) was going to require constant around-the-clock coordination.

On his desk was a dispatch from the First Submarine Flotilla, a unit outside of his command, sent to him as a courtesy. The *Perseus* had failed to make her last radio check. There were any number of reasons why that might have occurred. No cause for undue alarm.

Still . . .

Colonel John Randal knocked on his door even though it was open, "Can I have a word, sir?"

"Come in, Colonel."

Col. Randal said, "Terry's father wants the Lancelot Lancers to return to the UK to prepare for the invasion—expects him to resume command."

VAdm. Ransom said, "Once the Duke gets an idea, an exceedingly rare occurrence in my experience, there is no dissuading him. A dark day for Raiding Forces and for you in particular. You two have been through quite a lot together."

Col. Randal said, "My thoughts exactly. I'm considering bringing Major Stirling in as his replacement. Stephanie recommended him. I'd like you to sign off on who takes the assignment prior to, sir."

"Excellent choice, 'Right Man, Right Job'—as long, that is, as we keep him away from the demolitions locker."

VAdm. Ransom did not mention the *Perseus.*

LIEUTENANT GENERAL "GERONIMO" JOE MCKOY WAS THE next person Colonel John Randal spoke to about a replacement for Lieutenant Colonel Sir Terry "Zorro" Stone.

"Percy'll do you a good job, John. No problem. He ain't Zorro—but then who is?"

"That is a fact, General."

"You've got nothin' to worry about. Stephanie'll whip ol' Pyro into shape—won't lose a step around here. The Major'll be workin' for her in no time. We both know that."

Col. Randal said, "That is a fact."

"Me and Waldo's goin' to the Congo to check out LONG NECK," Lt. Gen. McKoy said. "That bad timin' with you headin' out to London?"

Col. Randal wondered how he could know about that. "Try to make it back before Christmas, General. I intend to return to ABC by then."

Lt. Gen. McKoy said, "I'll push our departure date up a little—that work?"

"Actually," Col. Randal said, "what I'd really like is for you to travel to the UK with me."

Lt. Gen. McKoy said, "Well, in that case we'll put off the Congo jaunt until right after Christmas. That way me and Waldo'll have time to make a detailed inspection a' Frank's operation without bein' in any big hurry to get back. I want to take a good long look at the whole shootin' match. Can't have any a' those sparklers he's confiscatin' down there fallin' through the cracks and endin' up in somebody's pocket."

Col. Randal said, "I'd like that."

"Anything you want to tell me about this England deal?"

"Not really. Just a feeling."

"Doesn't make me real comfortable when you get to havin' those kind a' thoughts, John."

"Probably nothing to it."

CAPTAIN BILLY JACK JAXX SPOTTED A HILL IN THE distance. There was a snow-white Greek Orthodox church on the crest. On Aegean islands churches were typically built on a high point overlooking the main village. The parishioners had to make a pilgrimage to reach it for services. The steeple would make the perfect observation point for his purposes. He decided to investigate. If any of the 1st Mountain Division Jägers came up from the village they would be easy to spot from a distance because there was only one narrow switchback trail to traverse. The terrain being rugged, the Germans would be channelized. He should have enough time to escape and evade (E&E) down the far side of the hill if any approached.

At least that was the plan.

The going was a rough slog. The hill, being a volcanic formation, as was often the case on the islands in the ATO, while not high was almost straight up. Capt. Jaxx took the precaution of circling around and making his approach from the land side—a really tough climb. While not able to see the village as he made his way up, he was pretty sure it would be located down below on the Aegean side.

Working his way to the top he came in behind the church building to screen his approach. The structure was built like a fort, constructed out of massive blocks of stone, some more than twice as tall as he was. Capt. Jaxx could not help but wonder how they had been carted up that hill back in the day.

There was no rear entrance. Possibly the thought, at the time the church was built, being that if the pirates who plied the Aegean were to raid the island the Greeks could retreat to the church and seek sanctuary by blocking the front entrance. Not the best of ideas since the buccaneers had probably heard that one before and besides, it worked to the pirates advantage. If the population ran away to hide, the invaders would be free to plunder the abandoned village at their leisure.

When Capt. Jaxx eased his way around the corner of the church to the front he came face to face with the abbot who was easily identifiable by the gold pectoral cross he was wearing around his neck. He was a wizened little white-haired gentleman attired in the traditional cassock of the Greek Orthodox Church. Suddenly encountering a heavily armed man in the uniform—except for the faded blue jeans—of a U.S. paratrooper was clearly a shock.

"Englisher?"

"American—any Nazis around here, Father?"

Capt. Jaxx was not exactly sure how to address this man of the cloth. He was no expert on the Greek Orthodox Church's chain of command.

In broken English the abbot said, "No, no . . . no Germans, my son."

If that were so, why had two Nazis armed with 9mm Steyr MP-34 submachine guns just exited the church and commenced fire at virtually point-blank range, spraying 9mm rounds from the hip? The Steyr MP-

34 has a cyclic rate of 550 rounds per minute. Multiplied times two, the Jäggers were putting out what was described in military terms as an intensive volume of fire. Their initial volley cut the abbot, standing between the alpine troopers and Capt. Jaxx, almost in half.

When the abbot went down—going slow in a hurry as advocated by Lieutenant General "Geronimo" Joe McKoy for dangerous encounters, Capt. Jaxx dropped both Nazis with a pair of crisp three-round bursts firing his captured Steyer MP-34 firing from the shoulder and calling his shots. The 1st Mountain Division Jägers may have been world-class climbers, skiers and hardened combat veterans, but like a lot of specialist units they focused on their unique skills and not individual shooting technique—counting on a heavy volume of fire to make up for marksmanship.

That was a mistake. Especially when going up against an experienced snap shooter who had grown up in rural West Texas handling firearms all his life. And who practiced a lot. Capt. Jaxx's grandfather, Sheriff Marlboro Jaxx, taught him from an early age "accuracy trumps volume of fire every time in a gunfight." Lt. Gen. McKoy frequently echoed a similar sentiment. "It ain't how fast you work the trigger . . . it's how quick you put rounds on the opposition."

Why would the abbot lie to him about there not being any Germans? Capt. Jaxx had no idea. Now he had two more submachine guns he did not much need. However, the extra ammunition in the dead Germans' leather M-34 magazine pouches was welcome. As were the four Model 24 potato masher grenades the Jägers had stuck in their belts.

Both Germans were equipped with canteens. Those he could use. Water might be a problem.

When no more Nazis showed themselves, Capt. Jaxx went inside the church to clear it. The place appeared empty. That could have been an illusion. As he passed the third level while climbing the bell tower, he could see what appeared to be cells with heavy round-topped wooden doors lining one side of the building. The doors were all shut. A monastery—there might be monks cowering inside.

The thought was a little unsettling.

The view from the top of the steeple was spectacular—though Capt. Jaxx realized he was violating the basic E&E principle of avoiding prominent terrain features. His thought was to observe the built-up area, gather intel, and then boogie. In the distance, the Aegean was a beautiful turquoise blue in contrast to the harsh arid terrain he had traversed from where he had washed up ashore. There was a fairly large village less than a mile down the hill. The houses sloped to the water's edge. A half dozen small caiques and a two-masted schooner were docked in the protected harbor.

He studied the layout through the 6x30 Zeiss field glasses he had liberated. On closer inspection the place looked like it had been a war zone in days past. Some of the structures appeared to have been bombed. More than a few were burned to the ground. The island had clearly experienced hard occupation under first the Italians and now the Germans.

Not a lot of people were moving around down in the village. The ones he saw were mostly women. A few fishermen were on the pier, but they were older men like the two who had led the Nazis to his cave. Nothing unusual about that. At the start of the war military-aged men left the islands to enlist in the Greek Armed Forces. Anyone else who could afford to had relocated to New Zealand, Australia, or Egypt.

Capt. Jaxx decided his initial estimate of a platoon-minus-sized occupation force was probably on the money. The Germans could not afford to station larger elements on islands that did not have military value. There was nothing he could observe that would indicate this place was anything more than a poor Greek fishing village of no strategic significance. Reflecting back on his Enemy Forces briefings there was another possibility to consider. And it was significant. On remote or unimportant islands, the Wehrmacht often placed a feldwebel in charge of understrength platoons or lesser-sized units assigned to occupation duty.

He might have already taken out the local enemy troop commander.

One building in town offered possibilities for being the 1st Mountain Division Detachment's Command Post (CP). It looked like the kind of commercial property that served as the local bank on other islands where Capt. Jaxx had come calling. Men in uniform entered and departed from time to time. The Gebirgsjägers always traveled in pairs—the buddy system. Likely having learned from previous campaigns it paid to beware of the locals in occupied territory. He tried counting but gave it up as a lost cause. At this distance there was no way to be sure he was seeing different individuals. Nevertheless, there were seldom more than six or eight enemy soldiers in sight at any given time.

The bank, if it was a bank, was not large enough to house a platoon-sized unit—even an understrength one. There must be a barracks somewhere. Capt. Jaxx was interested to know where that might be.

He intended to pay a visit come zero dark thirty, or maybe a little later.

COLONEL JOHN RANDAL AND MAJOR THE LADY JANE Seaborn were sitting alone at a table in the Other Ranks (OR) mess hall. After mealtimes it was open all day and could be used by anyone. A lot of Raiding Forces business was conducted there over tea or coffee.

Lady Jane was excited by the prospect of a trip home. Col. Randal was not. The idea of being away from Castelrozzo for an extended period of time to do who knew what during this period of reorganization was not something he wanted to do. He was trying not to show it.

Lady Jane said, "Bronc contacted me. He is sending his personal C-47 to fly our party to London. Asked me to inform you he would meet us there. Naturally, we shall be staying at the Bradford. I have arranged a room for him on my private floor—your old suite actually.

Col. Randal remembered it had also been her husband, Mallory's, before him.

Lady Jane said, "General Donovan is in residence there as well. With all the high-ranking Yanks arriving in-country, five-star hotel accommodations are almost as rare as unicorns. Especially those in the Mayfair district. The Bradford is fully booked, as is Claridge's and the Savoy.

Col. Randal said, "Wild Bill's in England?"

"He flew there from Cairo to meet with David Bruce, the OSS London Branch Chief, before returning to the States," Lady Jane said. "When the General learned of our trip he decided to postpone his departure until after your meeting with my godfather. He is most interested to learn how it turns out for reasons not known to me."

Col. Randal said, "Dudley Clarke set up the meeting. Didn't tell me what it's about either. I thought maybe you could."

Unable to restrain her excitement, Lady Jane laughed, "Who cares? We take a whirlwind vacation. You talk to Uncle Johnny about whatever—dying to show him my Sheba diamond. Then we come back to ABC to celebrate Christmas . . . fun!"

Col. Randal said, "So, am I supposed to ask Colonel Bevan for your hand in marriage?" He could not help flashing back to the mental image of Mallory's feet sticking up in the air from a latrine in some Japanese POW camp where he had been "boreholed" to death by his fellow prisoners.

Lady Jane laughed, "The only permission you need is mine, babe." She did not mention that when he had briefly met her godfather four years ago, it had not gone well. Uncle Johnny had been less than impressed with her boyfriend—he told her so.

Col. Randal said, "Good to know."

Lady Jane said, "Are you terribly disappointed Terry shall be leaving Raiding Forces? Not surprised actually. The Duke would want the family regiment to be in the van at the kill."

Avoiding the question, Col. Randal said, "Stephanie recommended Major Stirling as his replacement. What are your thoughts?"

Lady Jane said, "An endorsement from Steph is all one needs. Pearcy is an outstanding officer. Does a promotion come with the assignment?"

Col. Randal said, "Terry's bringing the Major in to understudy him. I'll wait to see how Pyro works out first. If he manages to blow up ABCHQ while we're away that would be a negative."

As usual when alone with her he started to unwind. Being around Lady Jane was addictive—like a drug. Col. Randal wondered, not for the first time, if she knew that.

There was no real reason for him not to go to England. He had devised an amalgamation plan involving selection and additional training for the new units being assigned or attached. "Assigned" meant the unit was a permanent part of Raiding Forces like OSS OG Det 3, 1st

Bn KO and the volunteer elements of 11th PB. "Attached" meant they could be transferred somewhere else at some later date, as was the case with the Small Boat Squadron (SBS) and Long Range Desert Group (LRDG). While the infusion of the new units was currently underway it would take time to complete. The process was in good hands. There was no need for him to be here.

Still in the "big picture," as the newsreels described what was taking place in the war, it was one thing to have developed a strategy like his Constant Pressure Concept (CPC)—another to implement it. He would have liked to see the process through. While CPC sounded simple enough, in fact it was a sophisticated combined operation involving air, sea and land components with heavy reliance on intelligence gathering and small unit leadership. Selecting the right people, putting the disparate pieces together so that Raiding Forces operated as a fine tuned team, not simply an aggregation of highly talented individuals was not like flipping a light switch.

All of Raiding Forces had to be prepped for the CPC mission. Supporting elements had to be put in place, a logistical plan to supply the schooners stationed along Turkey's coast developed, and the small-scale raids, which for the most part would be independent operations led by junior officers, gradually phased in until a full operational rhythm was achieved. A lot of teamwork both at the staff and small unit level was going to be required to carry it off. And both of those required constant command support and supervision.

By the time Col. Randal returned from England the SBS, 1st Bn KO, 11th PB and the amalgamation of the Red Indians of 30 Assault Unit and the OSS OG Det 3 with handpicked elements of Major Butch "Headhunter" Hoolihan's former troop of Royal Marines would be in full swing.

The LRDG, already pulling Beach Watch missions on islands scattered throughout the Aegean even while recruiting for additional LRDG reconnaissance operators to make up for the units' excessive losses on Leros, would have expanded the numbers of its patrols in the field. And more of the 575th Ranger Task Force personnel would have made their way back from behind the lines in Italy to rejoin.

Maybe this was as good a time as any for him to have a short standdown. Not that he had much choice. No one asked if he would *like* to travel to the UK.

Col. Randal said, "You need to let Fleming know we'll be in town."

Lady Jane laughed. "Absolutely, you two have to coordinate capturing the most beautiful woman in the world. One would not want to forget that."

Ignoring her, Col. Randal said, "Pencil General McKoy and Mr. Treywick in on your manifest. They'll be traveling with us."

Lady Jane said, "May I invite whoever I want?"

Col. Randal said, "Roger that—it's your trip. I'll need to meet with General Donovan, Colonel Bevan and Commander Fleming. I want to pay a call on Seaborn House. The rest of the time I'm available for your plans exclusively. Make it happen, Lady Jane."

"I love you, John Randal."

MAJOR THE LADY JANE SEABORN WENT TO SEE HER UNCLE, Vice Admiral Sir Randolph "Razor" Ransom, in his office. He was leafing through a stack of papers on his desk. Ever since her family had been killed in a plane crash in Kenya the Admiral had treated Lady Jane like a daughter.

"Would it be possible to recall Brandy and Parker to ABC?"

"Why do you ask?"

"John and I are flying out to London. My godfather, Uncle Johnny, has requested an in-person interview with him for some reason—Dudley stage-managed it apparently. I would love for Brandy to go, and Parker would be able to visit her husband at Seaborn House."

VAdm. Ransom said, "Consider it done, Jane. Anything for you. Along those lines possibly you can do something for me in return. There has been a late-breaking development in the command structure of light surface units in the Aegean. As of now they all fall under Small Raids Inc. We have inherited additional MGB and ML assets that shall prove invaluable for Raiding Forces and MI-9 missions.

"Which means we can reduce SRI dependence on the MAS boats. Do you have anything Brandy and Parker might do to keep them occupied? Those two are not going to take it well when I put them on the beach. Not looking forward to that conversation."

Lady Jane said, "We have started working up an anti-collaborator program—OPERATION BAD CALL. The idea is to snatch up Greeks collaborating with the Germans or, if necessary, land ashore and liquidate them. Brandy and Parker would be a perfect fit. Both are MI-5, speak Greek and could supply sea transport for our missions since the MAS boats will no longer be constantly at sea running errands for you."

VAdm. Ransom said, "You are a lifesaver, Jane! I shall detail the MAS boats to BAD CALL on a permanent basis. Brandy and Parker are off the SRI daily operational rotation. Effective immediately they work for you."

Lady Jane said, "Perfect."

VAdm Ransom said, "Actually it is, Jane. Your counterintelligence operation obtains the services of two experienced MI-5 agents with language capabilities and BAD CALL has organic sea transport. No more having to submit a request for transportation that might or might not be immediately available when you have the odd hasty mission crop up.

"I shall need to keep up with what your BAD CALL people are doing so SRI can be prepared to come to your support in the event that ever becomes necessary. Simply coordinate with my office the when and where—then go."

Lady Jane flashed one of her best-grade heart-attack smiles, "A pleasure doing business with you, Uncle Razor."

VAdm. Ransom said, "Always."

Next Lady Jane went in search of Mandy Paige. She found her coming out of Dr. Layton Winthrop's office with her mother, Veronica. MI-9 (Escape) and SOE both had agents scattered around the Aegean. While the two organizations had entirely different mission profiles they shared much in common. The problem was SOE had a reputation for infighting with other agencies. It saw Mrs. Paige's MI-9 organization, which was rapidly extending its tentacles throughout the islands and beyond to the mainland, as a competitor encroaching on their territory.

Veronica Paige maintained a small fleet of caiques commanded by Sergeant Major Mike "March or Die" Mikkalis and had her own agents installed on several of the enemy-occupied islands with plans for more. Her MI-9 had easy access to the entire ATO by being able to stage from the Raiding Forces sloops moored in remote out-of-the-way coves along the Turkish Coast—an asset SOE did not have access to.

Dr. Winthrop was an SOE agent handler working from Raiding Forces Headquarters on ABC running a network of archeologists/spies throughout the islands. At least that was the cover story SOE was led to believe. In fact, the Doctor had a longtime pre-war MI-6 (Secret Intelligence Service) relationship dating back to long before SOE had even been founded. In addition, the professor was now effectively functioning as Colonel John Randal's S2 (Intelligence Officer). He had an excellent relationship with Veronica and was one of, if not *the*, biggest supporter of her efforts to bring back Allied personnel trapped behind enemy lines. They freely shared intelligence information.

Mandy did not work for Dr. Winthrop or her mother, but she liked to keep her finger on the pulse of anything and everything involving Raiding Forces. She considered herself Col. Randal's personal liaison officer-at-large—to everything.

Lady Jane said, "Mandy, a moment?"

The two walked to the OR dining hall. When they arrived the Mess Sergeant rushed over to escort them to a table. He personally took their order for tea. Not normal practice. The OR mess was self-service regardless of rank or station in life—unless Lady Jane dropped in.

Lady Jane said, "John and I shall be flying out to London in the next day or so. Under normal circumstances we would love to invite you to travel with us. These are not ordinary times. I need you to remain here and supervise BAD CALL."

Mandy said, "Love to—my pleasure."

Lady Jane said, "I want you to be my deputy on BAD CALL. In the future, on occasion I may have to be away for extended periods. That means you shall be in charge."

Mandy said, "The two of us working together is a dream assignment."

Lady Jane said, "With Jack away at sea, which SOG officer do you want to have on standby in the event a mission presents itself?"

"Chase Starrett."

"Are you certain—he is awfully young?"

"If Billy Jack believes Chase is qualified to be an SOG officer, then snatching up some Greek collaborator from their bed during the middle of the night should be easy enough."

Lady Jane laughed. "Absolutely right. What am I worried about? He is only a few years younger than me. With everything that has happened in the war I do not feel so young at times."

Mandy said, "Well you are. Enjoy your visit home. I have this."

Next stop for Lady Jane was Captain Stephanie Fawcett-Tatum's desk. The two had been friends for ages. They attended the same exclusive Swiss finishing school, *Institut Le Rosey* aka "Rosey" to the girls.

"John and I shall be traveling back to the UK shortly."

"So I understand."

"What time is Beverly flying in?"

"Landed a few minutes ago. Should be walking into the TOC any moment now."

"When are King and Pam expected back from Morocco?"

"Tuesday."

Since Capt. Fawcett-Tatum was not read into LONG NECK, she had no idea what the purpose of the frequent jaunts to Morocco might be. As a cover story, King had mentioned he was attempting to establish a source of wristwatches out of Spain to purchase for the Royal Air Force. Which was not entirely untrue. While he had virtually cornered the Moroccan illicit diamond market trade—a top-secret project classified to anyone outside of LONG NECK/CARD GAME, the Merc was not having much luck monopolizing the colonies' watch trade.

Lady Jane said, "Are you able to contact them?"

Capt. Fawcett-Tatum said, "They are staying at the La Tour Hassan Palace. We can send a wire."

Lady Jane laughed. "King and Pam have quite the taste in hotels."

Capt. Fawcett-Tatum said, "Originally built for Sultan Moulay Hassan in the last century. I spent a week there before the war . . . fabulous! We should go sometime, Jane."

Lady Jane said, "When you compose the message, say, "URGENT STOP TIME SENSITIVE STOP IMMEDIATE RETURN ABC REQUESTED STOP" signed JS FOR JR. I want them to travel to London with our party."

Capt. Fawcett-Tatum said, "I shall arrange for the TWX to go out immediately. By the way, a letter from Major James Riddell arrived earlier today. You remember him—the good-looking vice-captain of the 1936 Olympic ski team. James is seeking qualified volunteers to be instructors at the 9th Army ski school he commands at Cedars—interested?"

Lady Jane laughed, "We should do it, Steph. Run away. Live on the ski slopes."

Capt. Fawcett-Tatum said, "In our dreams."

VERONICA PAIGE TRACKED COLONEL JOHN RANDAL TO the suite he shared with Major the Lady Jane Seaborn. Flanigan was on the desk. The two had been at Habbaniya prior to and during the siege making them card-carrying members of the Habbaniya Mafia. He waved her in.

Col. Randal was cleaning his pistols on a towel spread over the coffee table in the living room when she arrived. One glance at Veronica and he knew the faint hint of tranquility he was beginning to experience prior to the trip to London was about to hit a wall, crash and burn.

"John, we have a pilot down on Sikinos Island."

"Where's Sikinos?"

"Approximately 250 miles northwest of Castelrozzo. It's a small island . . . not sure of the details. One of the LRDG Coastwatch Teams—Baker Team 6, is on an unpopulated key two miles offshore. It reported that a Spitfire, believed to be a photo reconnaissance aircraft, flew over at low-level, trailing smoke. The pilot bailed out over Sikinos at 1345 hours today. In addition to notifying MI-9, the LRDG report was forwarded to the Northwest Africa Photo Reconnaissance Wing . . . a joint USAAF/RAF unit. It conducts photo recon in the Aegean. The NAPRW has requested we make every effort to retrieve their pilot. My guess would be he is in possession of classified information about the technical photographic equipment the Wing uses no one wants compromised."

Col. Randal said, "Wait one."

He picked up the phone and dialed Captain Stephanie Fawcett-Tatum's desk.

"Stephanie, have Lieutenant Starrett meet me in the TOC—Standby Ready."

"Aye, aye, sir!"

Standby Ready mean a mission was eminent the only question was when to move out.

"Tell the professor I'm on the way down with Veronica. We're going to need a briefing on Sikinos Island."

"Anything else, John?"

"Negative, we're moving now."

Lieutenant Chase Starrett was waiting at Capt. Fawcett-Tatum's desk when they arrived. She said, "Dr. Winthrop is available at your pleasure, John."

"Where's Major Stirling?"

"With his squadron off the coast of Turkey."

"Go ahead and notify him to report to me here—your new boss."

Col. Randal felt more than a little stupid saying that.

"Excellent choice, John."

"Glad you approve."

Capt. Fawcett-Tatum managed to maintain a perfectly straight face.

When Col. Randal, Veronica and Lt. Starrett walked into Doctor Layton Winthrop's office, Vice Admiral Sir Randolph "Razor" Ransom was there waiting. From across the TOC Lieutenant General "Geronimo" Joe McKoy, saw something was up and followed them in with James "Baldie" Taylor right behind, tailed by Mandy who did not want to miss out on anything. Lieutenant Colonel Sir Terry "Zorro" Stone came in right behind them.

Beverly Blackwell, arriving in the TOC with her flight bag over her shoulder, came over to see what was happening in case her services might be required. So many people crowded into the professor's office that everyone had to stand around his desk.

Dr. Winthrop's FANY, Lieutenant Violet Pelkington, a new addition to Raiding Forces, was charged with being the guardian of his door. She did not even bother trying to stop the late arrivals streaming past her desk. Besides, she was curious about all the excitement. And wondering what a dreamboat like Chase Starrett was doing in such a high-level conference.

Through the plate glass window, Lt. Pelkington could see Beverly standing next to Col. Randal bumping up against him from time to time in a casual friendly way. Col. Randal ignored her but did not seem to mind. *Hmmmm?*

Veronica went first, "The short version of what I am about to relate—there is no long version—is LRDG Team Baker-6 observed the pilot of a photo reconnaissance Spitfire bail out over Sikinos Island earlier today. Northwest Africa Photographic Reconnaissance Wing—a multinational composite unit consisting of a U.S. squadron, an RAF squadron, a South African squadron and a Free French squadron, was notified of the shootdown, as was I. The commanding officer of NAPRW has confirmed one of their aircraft is missing and formally requested MI-9 to recover their flyer. I informed Colonel Randal and here we all are."

She handed the TWX (Teletypewriter Exchange Service) she had received from the photo recon unit commanding officer to Col. Randal, who glanced at it and passed it to VAdm. Ransom, who studied it before handing it to Lt. Gen. McKoy.

VAdm. Ransom said, "What can you tell us about Sikinos, professor?"

Dr. Winthrop said, "It's located a little over 250 miles northwest of ABC approximately 150 miles from Athens on the Greek mainland. The island is less than 20 square miles in size with a population of fewer than 200 Greek islanders. Before the war an estimated 1,000 Greek Fascists and Socialists were deported to the island during the Greek dictatorship of Ioannis Metaxas. The newcomers outnumbered the locals. They are still marooned there.

"The name Sikinos dates from antiquity when the king of Lemnos had to escape to the island from his own because the women revolted and started killing the men one by one . . ."

Col. Randal said, "Any Nazis on the island?"

"Seventeen under a certain Sergeant Walther Bauer."

Lt. Gen. McKoy said, "Pretty specific answer there, Doc."

Dr. Winthrop said, "I have an agent on Sikinos."

VAdm. Ransom said, "Are you in radio contact with him?"

Dr. Winthrop said, "Negative. My operative sends letters written in invisible ink by caique to a neighboring island thirty miles distant where another of my people has a radio. Crude but effective spy craft. In the past we have never had a necessity for more timely communications. Nothing of any military significance has ever occurred on Sikinos. At least until now."

Col. Randal said, "What's the terrain like?"

"Rugged."

"Any place for a drop zone?"

"From a map study—I have never actually been to the island—there is a flat patch at the base of the hill outside Chora, the main village, that may possibly be suitable. I would not trust any other location on the island to be a DZ without onsite inspection in advance of a drop. The deportees had to farm to survive. They have terraced virtually every inch of the landscape. That makes it chancy for parachutists."

Veronica said, "If we contacted the LRDG team would it be possible for them to paddle over with a radio and link up with your agent to help coordinate a rescue?"

Dr. Winthrop said, "Be better if I went."

Col. Randal said, "Doctor, is there any question we should have asked you that we failed to ask?"

Dr. Winthrop said, "An astute question, Colonel . . . none that I can think of at the moment. I shall get back to you if one comes up."

Lt. Gen. McKoy was studying the TWX from the reconnaissance wing. "Anybody notice the signature on the NAPRW telex . . . E. Roosevelt, Colonel, USAAF?

"No? E. stands for Elliott, one a' FDR's four boys. What we have here is a bona fide official request from the son of a' sittin' president a' the United States, hisself. A bona fide high-profile mission requirin' a full-court press has just landed in our lap big time."

Col. Randal knew there was no delegating this one—trip or no trip.

3

THOSE PESKY MONKS

COLONEL JOHN RANDAL CALLED HIS KEY PERSONNEL together in the TOC, "Listen up, people. Starting now General McKoy will be honchoing mission planning. Everyone reports directly to him. When you have a question or something develops, see the General. I'll be available—you can always talk to me but go through him first to maintain chain-of-command.

"Set up in my office if that works for you, General."

"Can do, John."

"Admiral, I need you and Beverly to work out how we're going to transport Doctor Winthrop the 250 miles to the Baker Team 6 location tonight. I want to hear our options and your recommendations, sir. I'm also going to need your thoughts on the best way to extract the team we send in to Sikinos to rescue Colonel Roosevelt's pilot. And we'll also require a plan for the exfil of all parties to include the LRDG Beach Watch Baker Team 6."

VAdm. Ransom said, "Aye, aye."

Col. Randal said, "Doctor, the LRDG can paddle you over to Sikinos. Once there link up with your agent. First thing—make a survey of the area you believe might work as a possible drop zone. Advise ABCHQ on the suitability—we need that ASAP same night, early on for mission planning."

Dr. Winthrop said, "Understood."

Col. Randal said, "Second, pin down where the Spitfire pilot is located. If he's on the run, you need to track him down. If you can't we'll postpone the op until you can. We have to know where he'll be at a certain time and a certain place before our Quick Reaction Team can go in."

Doctor Clayton Winthrop said, "Wilco."

Col. Randal said, "Lieutenant Starrett, put together a roster of available personnel here on ABC we can draw from to organize the QRT. Provide the list to the General as soon as you have it. I want to see a copy."

"Yes, sir."

Col. Randal said, "Beverly, make sure the new Hudson is rigged for a jump in the event we decide to make one. We need to know the current status of our duty pilots. Find out who's available or who we should bring in to fly the mission. Provide your findings to the General then come locate me—I want an in-person report."

"On it, John."

Col. Randal said, "Jim, an intel summary from your sources on recent enemy activity on the islands in the Sikinos area would be useful."

"I shall see what intelligence is available," Jim said. "Don't expect much. I am not personally aware our side has had a reason to do anything in that particular area up to now."

Col. Randal said, "Veronica, as soon as you have a moment after we break here I need to speak to you.

"OK then, hold your questions for the General. He'll meet with you individually. I say again—everything flows through him. Let's do this."

As he headed to take up residence in Col. Randal's office, Lt. Gen. McKoy said, "Sorta failed to mention 'Actions On the Objective,' John. Ain't your style to be leavin' out the single most important element."

"You and I handle that," Col. Randal said.

CAPTAIN BILLY JACK JAXX WAS SNEAKING ALONG THE coastline staying to the rocks and scrub brush running parallel fifty yards or so inland. This was a rugged island. The difficult terrain started right off the high-water mark. From his perch in the church bell tower he had observed two 1st Mountain Division Jägers swimming at an isolated beach half a mile east of the village. It seemed like an open invitation to pay them a call.

By the time he arrived near where he had seen the Germans the two men had returned to the spot on the beach where they had left their weapons, web gear and towels and were sunbathing nude. Capt. Jaxx wondered what that was about. Possibly he was reading too much into the situation—maybe it was a European thing.

Being careful not to make a sound, Capt. Jaxx checked the action on the Steyr MP-34 to be absolutely certain a 9mm round was in the chamber. There was. Just like the last three times he had checked. There is no such thing as being too sure. The 9mm Steyr SMG—not to be confused with the 9mm Bergmann MP-34 SMG or the 9mm Erma MP-34 SMG—was loaded. Wehrmacht weapons nomenclature could be confusing. The Steyr MP-34—though an older design—was, according to Lieutenant General "Geronimo" Joe McKoy, arguably the finest submachine gun in the German Army inventory. It was issued to elite units like the Waffen SS, Brandenburger Sea Raiders, or 22nd Air Landing Division.

Captain Roy Kidd, the Raiding Forces weapons expert, concurred with the General's assessment of the Steyr SMG, as did King, who pointed out it was expensive to manufacture. That said, Capt. Jaxx noted none of them carried one as their primary weapon. Raiding Forces trained extensively with enemy weapons in order to be prepared for circumstances such as the one he found himself in now, but no matter how much he practiced, he could never quite get used to the side-mounted magazine. You had to be careful to resist the temptation to hold the stick magazine with your left non-firing hand to prevent the weapon from tilting over to the left—that tended to cause the Steyr MP-34 to malfunction.

The two Germans were stretched out on their backs tanning. Gazing out to sea. Capt. Jaxx stepped out of the rocks and onto the soft sand. He advanced to within five feet of where the Nazis were reclining making sure to stay out of the sun to avoid casting a shadow. The Jägers never noticed him, possibly distracted.

PAP, PAP—PAP, PAP.

With the selector switch on semiauto Capt. Jaxx shot them both—two rounds each, center of mass upper body—fast. The Nazis flinched but neither cried out or made any other sound. They were dead, or at least comatose preparatory to being dead . . . or so it seemed. Instinctively he wanted to put another round in each of them to make sure but resisted the temptation. Protocol for 9mm and lower calibers was "two to the body one to the head."

On semiauto the MP-34 was not silent but it was substantially quieter than the 9mm Walther P-38 would have been. Barrel length makes a difference in sound signature. Still, the gunshots had been fairly loud. This far from the village and with the breeze gusting in from the sea dissipating the sound, it was doubtful anyone had been alarmed by the shots or would be by two more. Nevertheless, as was commonly said in Raiding Forces, "Why take a chance?"

Usually that sentiment called for a little more firepower to be placed on target—but not this time.

Capt. Jaxx surveyed the Nazis' gear. Both men were armed with 7.92mm Karabiner 98k rifles which were of no interest to him. He threw them into the water as far out as he could. Each had two stick grenades, which he added to his growing arsenal of potato mashers. And there were the 1st Mountain Division knives that he clipped on his bandoliers—he might be overdoing it with all the blades but they were pretty neat. Besides he had plans for them.

The phrase, "Why take a chance?"—one of several unofficial Raiding Forces mottos—kept going round and round in his head like a broken record.

Capt. Jaxx knew the best military outfits had their own standard operating procedures (SOP). Some in the form of mottos. These might seem sophomoric to an outsider. But oft-repeated unit-specific SOP pearls of wisdom are the secret sauce when the commander and/or his subordinate leaders have to make tactical decisions under duress and time is short. Even the Boy Scouts say, "Be Prepared." There was not anything wrong with that one. It might have been influenced by Major Robert Rogers' Rangers motto, "Don't Forget Nothing"—also frequently repeated in Raiding Forces. No one knew for sure if the Major ever actually said that or if MGM made it up for the movie *Northwest Passage* about Rogers' Rangers.

Capt. Jaxx was of the opinion that if Major Rogers did not say "Don't Forget Nothing" he should have.

When faced with a range of options, at times none of them good, standing orders in the form of rules, unwritten mottos and unit sayings showed the officers, NCOs and men of Raiding Forces how to approach the decision-making process. Everyone knew exactly what was expected of them. The goal was to achieve the best possible outcome.

This sounds obvious in calm circumstances but it sometimes gets diffused in the process of preparing to do bad things to bad people or trying to save yourself. Forward-leaning jump-ahead planning in the form of SOPs, rules, mottos, etc., is priceless when a leader is faced with making a tactical decision on the ground in the heat of the moment—the result of which will determine success or failure of the mission and/or possible life or death for himself and/or his troops.

There was something else. While attending British Commando School at Achnacarry, Scotland, Capt. Jaxx was taught when operating behind enemy lines in the event of discovery by anyone who posed a risk of giving away himself or his team's presence to the enemy they were to be eliminated with extreme prejudice. A polite way of saying *killed* without actually having to say it, on the grounds that "dead men tell no tales." It was advice straight out of *Treasure Island,* written in 1883 but applicable to this day. In fact, to this very minute.

Military decision-making in a hostile environment is not as easy as it sounds. Not everyone can do it. Especially when the situation is confusing. Right now Jack Cool was not entirely sure the two sunbathers were dead. While they appeared to be KIA there was no guarantee of that and he was not a medical professional able to determine so with an absolute certainty. The last thing he needed was for one of the nudists to come to life like in the cowboy movies and say, "They went "that a' way.""

"What the hell," Capt. Jaxx thought to himself. Like Waldo always says, "the problem is the solution.'"

A big advocate of Raiding Forces Rule Number 1, "The first rule is there ain't no rules," Capt. Jaxx took out one of the special issue 1st Mountain Division knives and slit both the Nazi's throats—he was nothing if not decisive. Besides Jack Cool had already burned his bridges when he shot those first three Germans in the back of the head.

Like Capt. Jaxx had concluded earlier. Poor prior planning produces poor results. Maybe he really should have thought that through a little better.

LIEUTENANT CHASE STARRETT HANDED COLONEL JOHN Randal a copy of the roster he had compiled of operators available to form a Quick Reaction Team (QRT). It was not encouraging. There were not enough qualified personnel on Advanced Base Castelrozzo to make up a team capable of taking on the seventeen Nazis Doctor Layton Winthrop said were on Sikinos. The list included the six instructors conducting the SBS officer selection course, the Lovat Scouts and Private Norvel "Horn Dog" Hansen. The rest of Raiding Forces personnel who would have normally been on ABC, like the Special Operations Group, were deployed elsewhere conducting training, being trained, on operations, or on leave incidental to Escaping & Evading from Italy following the widely scattered Benevento jump.

The Constant Pressure Concept (CPC) required every available man for the round-the-clock pinprick raiding program. Teams operating from Levant Schooner Flotilla caiques based along the Turkish coastline were hitting targets somewhere every night. It was said the Germans did not like to fight at night. That reluctance—when combined with the element of surprise achieved by arriving out of the dark from the sea unannounced, unexpected and concentrated at the point of attack—gave Raiding Forces an edge despite their lack of numbers.

Col. Randal was personally not convinced about the Germans' unwillingness to fight at night but he intended to conduct as many raids as possible under cover of darkness. His troops trained for nocturnal operations. They were good at them.

However, this mission to rescue the NAPRW pilot posed a number of problems. To have any chance of success, intelligence pinpointing the exact location of the airman had to be obtained. Was he on the run, in hiding, being held by the Germans? If so, how many were there, what was their unit? The caliber of troops Raiding Forces would be going up against was another piece of information crucial to mission planning. Professor Winthrop had not said and Col. Randal had failed to inquire.

He picked up the phone on the desk he was standing next to and asked the switchboard operator to put him through to the professor. "I forgot to ask who we're going up against on Sikinos, Doctor."

"Our old friends—the 999th."

Good news and bad news. Low-grade troops but they murdered people for sport. Bringing back a dead West Africa Photo Reconnaissance Wing pilot for the President's favorite son would not be good. Right about now he could use one of The Great Teddy's magic tricks—Hey, Presto!

"You don't have Lieutenant Hamilton on your roster," Col. Randal said to Lt. Starrett.

"I only listed qualified operators, sir—he's our camouflage officer."

Col. Randal said, "Put him on there. Teddy's gone to all the schools and he's been a strap hanger on more missions than I can remember—pretty good with my Brixia 45mm too. We're going to need every trigger puller we can get."

"Can do, sir."

"I want you to serve as my aide until we work out troop assignments, Lieutenant Starrett. Don't worry, stud, you'll be leading a team when we go in."

"Yes, sir!"

"One more thing, see what it's going to take to get Captain Hays, Captain Bonham, Captain Novak and Lieutenant Mascuch to ABC. If it's at all possible I want them here ASAP. Also, I need the GSS interpreter Xanthos."

"Aren't Clint and Jake lieutenants, sir?"

"Not anymore. We need troop commanders. Lieutenant Mascuch will be promoted too as soon as he gets a few more ops under his belt. You're going to need to step it up. Your services are about to be much in demand around here, Lieutenant."

"Yes, sir!"

At this point Col. Randal knew he needed to stand back and let his people work—a lot easier said than done. He wanted to be in the thick of the mission planning process. Although what was being contemplated was a tiny operation, in scale it was an extremely complex Combined Operation with air, land and sea components, carried out at long range with no possibility of support in the event things went wrong.

And as he was now fully aware, Raiding Forces did not have sufficient troops immediately available to execute it properly.

Lieutenant General "Geronimo" Joe McKoy spotted Col. Randal talking to Lt. Starrett and came over, "I just got a chance to take a look at your roster of available troops, Chase. Include me and Waldo. We ain't sittin' this one out."

"Yes, sir."

Col. Randal started to object but stopped himself.

Major the Lady Jane Seaborn walked into the TOC and dropped by Captain Stephanie Fawcett-Tatum's desk. "John and I shall be in the Other Ranks dining hall if anyone should have need of either of us."

Knowing him as well as she did, Lady Jane was confident Col. Randal could use her company about now. He hated the idea of standing aside while others went about developing enough information to work up the initial concept of the operation. Her plan was to keep him occupied and out of everyone's hair while they did their jobs.

VICE ADMIRAL SIR RANDOLPH "RAZOR" RANSOM AND Beverly Blackwell were in the TOC studying the giant wall map of the Aegean Theatre of Operations (ATO). Colonel John Randal joined them. Everyone in the room was keeping a discreet eye on them. No one would come anywhere near. The three were engaged in a strategy session with enormous repercussions for the mission ahead. It was not lost on the people in the TOC that VAdm. Ransom and Col. Randal treated the former Texas beauty queen as a colleague.

Age, gender, even rank mattered little in Raiding Forces. Respect was earned, not given. Competence backed up by performance was the measure—nothing else mattered.

Sikinos was one of the lesser inhabited islands in the Cyclades chain. It only had one village. The terrain was extremely rugged in a small way. Flying time from Castelrozzo was a little over an hour and a half. By boat it was eight to ten hours depending on the class of the craft making the run.

Beverly said, “It’s not a difficult mission to fly, Admiral. We’ll go in at low-level partway to avoid radar on Rhodes. It’s still not confirmed that the Luftwaffe has night fighters stationed on the island. So we’ll have to craft a flight plan that presumes they do.

“The problem is there’s no place on Sikinos to drop the QRT. I don’t see a single location I’d recommend that’s large enough for a DZ. The last thing we want is injuries on the jump to hobble the assault team. My recommendation is to scratch the idea of a drop.”

VAdm. Ransom said, “Colonel?”

Col. Randal said, “If Beverly advises against a parachute insertion on Sikinos then there’s no jump, Admiral.”

Beverly glanced sideways at Col. Randal. He had given up a little too easily. Did he know something?

The village, whose name was “Village” in Greek, was located on an escarpment rising from the Aegean with snow-white houses, at least the ones not burned out, flowing down to a small harbor. There was the typical Greek Orthodox Church on the crest of the hill high above it. Even if they could locate a good place to drop in, the raiding party would still have to scale the hill, bypass the church, work its way down the escarpment, then infiltrate through the village to get at the 999th Light Division detachment. Since they would be making a night drop likely the Criminals would be found in a tavern they had requisitioned, drinking themselves blind drunk, as was the unit’s customary method of operation when conducting occupation duty on a remote Aegean island. Meaning, basically, only being there to show the flag.

At this point the location of the German remain overnight position (RON) had not been established. And with the primitive state of the radio communications network in the ATO, it would not be until Doctor Layton Winthrop could land ashore on Sikinos and make that determination. He would have to discover the 999th Light Division troops' RON—meaning tavern—then radio that information back to the Long Range Desert Group Beach Watch Team Baker 6. The LRDG would forward the message to one of the two Small Raids Inc. H-Class submarines standing by to relay to Advanced Base Castelrozzo, where it would be transmitted to the airborne Hudson transporting the raiding party to the island. Only then could a final plan be crafted to locate the NWAPW pilot and bring him off the island. This operation was taking the military term "hasty" mission to a level never imagined when the definition was originally created by the Infantry School's Tactics Department at Ft. Benning, Georgia.

VAdm. Ransom said, "First things first. How do we transport the professor the 250 miles to the Baker 6 location, Beverly? If by boat, in the event we were to depart immediately, the doctor would still not be able to link up with the LRDG, be paddled over to Sikinos by rubber assault raft, gather intelligence on the pilot's location and radio it back in time for us to launch the QRT before BMNT tomorrow."

Beverly said, "Johnny doesn't like daylight raids."

VAdm. Ransom said, "No, he does not."

Neither VAdm. Ransom nor Col. Randal failed to notice Beverly had not answered the question.

Col. Randal was standing right there observing while they were talking about him but did not insert himself into their conversation—he knew exactly what to do.

Major the Lady Jane Seaborn came to retrieve him. "John, your presence is required in the Other Ranks dining hall—follow me."

CAPTAIN BILLY JACK JAXX WORKED HIS WAY BACK uphill. He needed to utilize the elevation of the church bell tower again in order to scope out the built-up area. It was a chancy move. Nevertheless, he considered it worth the risk given the opportunity to study the village, assess the German reactions to him taking out seven of their troops and plan his next move. Moving through the rocky terrain he maintained a line of sight on the winding path up the steep incline from the village. If any Nazis were traveling along it to check on the two Gebirgsjägers he had shot at the church earlier, he wanted to know.

Jack Cool was not in the mood for any surprises when he reached the top.

At the military crest—a term describing the terrain just below the actual top of a mountain, hill or ridgeline because you could move along it and not be skylighted—he went to ground to study the situation. Taking his time, Capt. Jaxx broke out the Zeiss 6x30s binoculars and glassed the grounds around the building. Things were not as he left them. The two 1st Mountain Division men were still where they had fallen, dead on the ground. But the abbot was missing.

Why might that be?

The only explanation he could come up with was the rooms on the third level were, in fact, cells. There *had* been people inside. Those pesky monks—or maybe they were nuns—must have snuck out after he departed the area and recovered the abbot's body.

Capt. Jaxx decided to continue the mission. He made a point of being as noisy as possible when entering the church to alert whoever was inside of his presence. And he moved slow to give them time to scurry back to their rooms. The dead abbot was collateral damage whose death he was powerless to have prevented. Still Capt. Jaxx felt bad about him—pretty sure the monks or nuns cloistered in the church did too.

He was nearly to the top of the stairway after it was too late to reconsider his course of action when Capt. Jaxx had a thought. Uh-oh! Would they blame him? There were not any monks and very few nuns in West Texas where he grew up so what he knew about monkhood amounted to exactly zero. Was there such a thing as warrior monks?

COLONEL JOHN RANDAL AND MAJOR THE LADY JANE Seaborn were at a table in the Other Ranks mess. The room was empty. In fact, Advanced Base Castelrozzo Headquarters seemed deserted with so many people away. Lady Jane was entertaining him with details of her plans for the trip to London. The idea being to keep him out of the TOC so the staff could do their workup on the mission without his looking over their shoulder. Because of her station in life Lady Jane was normally reserved around anyone outside a small circle of people she trusted. When it was just the two of them alone she talked all the time.

Though he was enjoying listening to her, Col. Randal was only halfway paying attention. Spending time alone with Lady Jane while she distracted him with matters that were not life-and-death was at the top of his list of favorite things. However, while she was going on about their agenda in London he was mentally running through a checklist of obstacles that had to be overcome for the rescue mission to have any possibility of success.

Extracting someone on the run E&Eing behind enemy lines or making prisoner rescues are the most difficult small-scale operations to pull off with no near second place. There were several reasons for the level of difficulty. The target's extraction must be carried out with surgical precision—something almost as rare in the military as "military intelligence".

When contemplating rescuing a high-value target behind enemy lines or carrying out a prisoner rescue, it is crucial to remember it's a flunk if the party being extracted is killed in the attempt. The target is designated as "precious cargo" for a reason. For there to be any chance of success, detailed target-specific information is a must. Knowing who, what, when, where and how many enemy personnel are physically present in the immediate vicinity of the person being rescued is an absolute requirement. In the case of the NAPRW pilot they were going after, worst-case scenario would be a Nazi holding a pistol to his head when the cavalry arrived—Col. Randal was going to need a contingency plan in case that happened.

Rescue missions require detailed prior planning and time to rehearse over and over and over again. Even with the best intelligence and all the time in the world to prepare precious cargo recovery operations have a low probability of successful resolution. Col. Randal did not have the luxury of either of those two vital preconditions—intel or time.

Vice Admiral Sir Randolph "Razor" Ransom and Beverly walked into the Other Ranks mess and headed to the table. Lieutenant General "Geronimo" Joe McKoy was trailing behind them. This was not a social call.

VAdm. Ransom said, "We have three immediate problems to resolve, Colonel. One, there are not sufficient troops here on ABC available to assault—with overwhelming force, the German garrison known to be on Sikinos. Attacking forces require at a *minimum* a three-to-one advantage. Two, the island cannot be reached from Alexandria, Turkey or ABC by sea in the time frame you have laid out, meaning tonight. That rules out a seaborne assault. Three, Beverly can fly Doctor Winthrop in to link up with the LRDG but then there is not enough turnaround time for her to return to ABC to pick up whatever scratch team you can cobble together and fly back to Sikinos. Which rules out a forced entry parachute assault—not that it matters as she's already vetoed that as an option."

Beverly said, "The island is pretty much equidistant from everywhere. Arranging air transport in the time window we have to work with is virtually impossible. To complicate things, the *Black Cat* is being serviced in Cyprus . . . an engine overhaul. We only have the Hudson and a pair of Walruses that are too slow to be much help. We've got a serious 'getting there' problem, Johnny."

Col. Randal said, "Here's what we're going to do."

He placed his coffee cup on the table. "This is ABC." Then he moved the saucer to the middle of the table. "This is Sikinos." Next he set a salt shaker close to the saucer. "This is the LRDG Baker 6's Beach Watch location.

"Admiral, are any of your new Coastal Forces Command boats currently operating in this expanse of the Aegean within a hundred miles or less of Sikinos?"

VAdm. Ransom said, "Affirmative—we maintain a screen of forward Air-Sea Rescue High-Speed Launches in the event one of our planes has to ditch at sea. An HSL is a sixty-three-foot craft designed by the British Power Boat Company. Essentially that is exactly what it is

. . . a lightly armed speed boat capable of forty-plus knots. If an HSL is not currently in the vicinity you require, I can have one in place within the hour."

Col. Randal said, "In that case, I'll order the OSS Operational Swimmers flown in from No. 4 Middle East Training School at Kabrit and dropped here on ABC. They'll load out on the Hudson along with the professor. Beverly, you'll fly them to rendezvous with the Admiral's Air-Sea Rescue boat. They'll transfer to the HSL at sea.

"Concurrently Dr. Winthrop will be flown to LRDG Baker Team 6 aboard a Walrus by one of our Special Operations pilots.

"The HSL will deliver the Det 3 team to Sikinos. They'll land ashore, neutralize the 999th and effect the rescue of the pilot based on the intel the doctor develops once he's on the ground. Meanwhile, the Walrus will be standing by to extract the LRDG and Dr. Winthrop's party when he pulls out.

"Comments?"

VAdm. Ransom said, "It's a difficult flight plan over enemy-contested water, quite possibly at least partially at night and requiring precision pinpoint navigation to locate a sixty-three-foot target. If Beverly feels she is capable of flying the mission, I can have a High-Speed Launch on station seventy-five miles off Sikinos standing by to take on the Frogs from the Hudson. The HSL will be capable of transporting them to Sikinos within two and a half hours from the time she delivers the team—possibly less."

"Sign me up I'll fly the mission, Admiral," Beverly said. "Provided you'll allow the HSL to turn on its lights to guide on when we arrive in the immediate vicinity if that should become necessary."

"Can do."

Lady Jane said, "The OSS lads shall have completed less than one full day of ground training, John, and you are asking them to parachute onto ABC?"

Col. Randal said, "Yeah, well, the Frogs have to make their first jump sometime—that's why the Airborne's preferred method of movement is called rapid deployment."

"Sounds like a rapid catastrophe," Lady Jane said.

CAPTAIN BILLY JACK JAXX WAS GLASSING THE VILLAGE from the bell tower. While he had taken out seven of what he estimated to be fourteen-plus men in the German occupation unit, there did not seem to be any sign of alarm among the Nazis he observed strolling around in pairs—the Jägers were religious about sticking to the buddy system. They viewed the threat the Greek civilians posed to Germans traveling alone as a very real danger and rightfully so. He could not understand why there was not any indication of an alert. Had the men he shot not been missed?

Being stationed on a small Aegean island and left to your own devices had to be boring duty after the newness wore off. Since the detachment had likely experienced nothing threatening in the way of Allied action since they arrived Capt. Jaxx could understand how the Germans might have stacked arms, gone their own way and not noticed anything was amiss. They probably believed the unaccounted-for men were off drinking in one of the small taverns, fishing somewhere, exploring the island or discreetly ensconced with a lonely Greek woman who did not want her neighbors to know she was giving comfort to the enemy.

This was especially possible if the feldwebel he shot was the NCOIC of the detachment on the island and not around to count noses or dream up work details to keep his men occupied. 1st Mountain Division was one of those squared-away outfits with sharp corners. In fact, now that he thought about it, it actually did look like no one was running the show down there.

Capt. Jaxx began to consider the possibility. When it began to get dark the Germans' laid-back attitude was going to change. No matter how peaceful things were right now, once nightfall began to set in, the shadows were going to become threatening and the Gebirgsjägers would no longer exhibit a tourist's mentality—not these veteran troops. There may not have been many Greek men left on the island and most of those were old but they fished for a living and were skilled in the use of the razor-sharp knives they all carried on their belts. The local fishermen were not lonely and had no interest in giving aid or comfort to their enemy. They wanted them dead—at least most did. Capt. Jaxx still had not figured out why some seemed to be collaborating.

As night fell the Germans would gradually begin to concentrate on self-preservation most likely in whatever location served as their barracks—at least initially. Then, if true to pattern the troops would gravitate to the bar it had commandeered for their personal use. No Greeks were allowed except for the women forced to provide entertainment. They would feel safe there. But at some point, even with no one in charge, the Nazis were going to notice a lot of their people were missing. Probably more than would be shacked up in the village somewhere.

Capt. Jaxx could see two ways to play the hand he had been dealt. He could wait until dark, trail the Germans to their bar of choice and attack them with the grenades he had acquired—a variation of the way Raiding Forces had done successfully on other occasions. That tactic would require him to hole up someplace and wait for the sun to go down. He would probably want to hit the bar at around 2400 hours to give the Nazis plenty of time to be drunk.

Or he could continue to pick off the Germans in out-of-the-way places when and where he could take them by surprise. This was the riskier of the two options but it would not require sitting around waiting.

Naturally, Jack Cool went with the second option.

LIEUTENANT GENERAL "GERONIMO" JOE MCKOY SAID, "That plan a' yours just may be the worst I've ever heard, John."

Colonel John Randal said, "I'm pretty sure you've told me that on several other occasions."

Lt. Gen. McKoy said, "You're probably right but this one surely takes the cake. There's about a hundred things that can go wrong and only one that can go right. "A lot a' rough edges on the concept a' the operation you laid out for the Razor—wouldn't really know where to start listin' my objections. Except to point out the 'rescuin' the pilot' part was pretty vague. Oh yeah, and you sorta conveniently skipped over the textbook requirement of needin' a three-to-one advantage for attacking forces. Had to have a sailor point that out."

Col. Randal said, "That's what I've got you for, General. To come up with solutions. As far as doing things by the book, I don't believe Raiding Forces has ever had three-to-one odds."

Lt. Gen. McKoy said, "Let's let things simmer for a little while. Give ourselves time to live with the idea a' how this mission needs to play itself out and see what develops. No tellin' what's gonna jump out to bite us next—first reports from the battlefield never bein' accurate.

"Same applies to initial intel reports."

Col. Randal said, "See . . . you're already off to a good start."

Lt. Gen. McKoy said, "You and me'll sit down somewhere and start whippin' this mission into shape later. Can't say it won't be interestin'. Go rescue a shot-down pilot somewhere on a nineteen-square mile island in the dark a' night two hundred fifty miles from ABC and do it right now with not enough people."

Col. Randal said, "I'd like to fly to Kabrit, link up with the OSS people and jump in with 'em back here."

Lt. Gen. McKoy said, "I can see how you would wantin' to lead from the front like you do. Those boys'll barely know how to put on a parachute. But I'm goin' to need you here, John. That trip would take you outta the loop for at least five hours and we ain't got five hours to give you to set the example."

Col. Randal said, "I was afraid you'd say that."

Lt. Gen. McKoy said, "Now you need to understand there's more to this mission than meets the eye. Lot a' ins and outs. Major politics are at play. We want to get it right but we also need to understand what's goin' on behind the scenes in order to look out for our own best interests."

"What are your thoughts, General?"

"We've got the President's boy Elliott—some say his favorite—in the mix. Always been the controversial son hangin' out with Howard Hughes in Hollywood, ridin' in big limousines, stayin' in fancy hotels, runnin' with movie stars and sportin' high-dollar hookers. FDR's always havin' to cover for him. Took Elliott to the Atlantic Charter meetin' in Argentia, then appointed him his military envoy for the Casablanca, Cairo and Tehran conferences. Makes sure everybody knows he's his son. The big generals don't like Elliott bein' involved but there's nothin' they can do, him bein' the President's son. Tends to rub their noses in his family connection. People I know ain't real fond a' him on a personal level, either, but what can they do?

"Then we've got the political climate in the Aegean to consider—Churchill's pet project. Marshall and Eisenhower . . . they don't want anything to do with it. Claim it'll suck men and equipment down a rabbit hole they'll need for the big invasion a' Europe. So the two a' them refuse to provide any support for the ATO. Word is Eisenhower's tryin' to demonstrate he can stand up to the Prime Minister… not giving the theatre any troops or military stores, knowin' it's his baby. There's those who claim that's due to all the generals he's passed over complainin' he only got his job because the Prime Minister handpicked him. There's those in the British camp who believed he'd be easy to bring around to the British point a' view on how to fight the war."

Col. Randal said, "I've heard that."

Lt. Gen. McKoy said, "So now the President's kid is a USAAF colonel in command of a joint multinational photo reconnaissance unit. And one a' his flyboys, who's apparently in possession of a lot a' Top Secret technical photo recon type information we don't want to fall into Nazi hands, is down on the run on an island in a theatre Eisenhower has refused to provide support for.

"Ain't real good for Ike, career-wise."

Col. Randal said, "I can see how that might not be."

Lt. Gen. McKoy said, "Then you got Donovan who the regular army pretty much hates for bein' outside the regular chain of command and for recruitin' their best men. Plus he's a personal friend a' the President, bein' his college football hero at Columbia—though Wild Bill told me confidentially he ain't real sure he ever actually met the man when they was in school. Now Donovan's in a position to ride to the rescue—that'd be us doin' the ridin' part.

"Churchill, he's most likely sittin' back watchin' this roll out, lickin' his chops. If it all goes wrong—like it probably will, he's in a position to blame Eisenhower for not backin' British efforts out here in the first place. And if it goes right, then he just might be able to leverage a little more theatre support from the U.S. for a job well done helpin' out the President's favorite son and all. Who, accordin' to one a' my old huntin' buddies, Churchill happens to hate—got drunk at the Tehran Conference a couple a' weeks ago an' sided with Stalin against him about shootin' 50,000 German officers after the war.

"Elliott was all for it."

Col. Randal said, "Really?"

"Oh yeah, caused a big stink."

"Never heard that story."

"Now, as for Wild Bill," Lt. Gen. McKoy said, "if we somehow manage to rescue this pilot for FDR, he's the man a' the hour—pure-D golden."

Col. Randal said, "What about Raiding Forces?"

Lt. Gen. McKoy said, "We were just grunts doin' our job."

"Lovely."

CAPTAIN BILLY JACK JAXX GLASSED THE VILLAGE LONG and hard. Not much of actionable intelligence was to be gained. Troops from the 1st Mountain Division were only occasionally to be seen. He seldom saw more than four German soldiers moving around at any one time and never all together in a group. There was nothing unusual about that. Experienced soldiers know to stay out of sight as much as possible when in garrison. It's the military version of hide and seek played in all armies all over the world. The rules are simple. If your sergeant cannot see you he has no way to place you on a work detail.

The Gebirgsjägers Capt. Jaxx spotted did not seem to be congregating in any one place, though from time to time some of them would show up at the small building he had tentatively identified as the detachment's Command Post. They never stayed inside long before departing on some errand or the other. There was no visible sign of alarm among the Germans. For all intents and purposes it was a lazy Aegean day. Occupation duty in the ATO was an easy assignment but about as exciting as watching paint dry—like being in an open-air prison in paradise. Except for the women there were those.

Capt. Jaxx was unable to establish a pattern of activity. Still a lot unknowns. How many enemy personnel were in the village was unclear. Why the Greek fishermen had led the Germans to where he washed ashore continued to be a mystery. As was the reason the abbot had lied about there being no Nazis at the church. Was the whole island collaborating with the enemy? If so it would be a first since Raiding Forces commenced operations in the ATO—Capt. Jaxx did not believe it to be the case.

Nevertheless, until proven differently, trusting the locals was not an option. Capt. Jaxx had no intention of making contact with them like he would have normally done. That was too bad. Greek women were widely known for their beauty. And with the military-aged men all gone to the Greek Army there had to be a lot of lonely ladies in the village. This place should have been a happy hunting ground. Some unlucky local girl was missing out on the joys of hiding him from the clutches of the evil Nazis all because he could not be certain of the islander's loyalties.

Capt. Jaxx might have been Escaping & Evading but there was no rule that said he had to give up his love life while doing it. Unless as in this case, there was the possibility the woman and or women might give him up to the Germans. That took the fun out of it. Or maybe not; he was still considering his options.

Finally Capt. Jaxx grew tired of observing nothing happening. By nature he was not a reconnaissance operator. That requires a particular mindset. While good at "sneaking and peeking" it was not his military skill of choice. It was time to take action. So, he went down the stairs of the bell tower, walked through the church and out the front door.

The two dead Nazis were still lying where they fell. He had given each of them a three-round burst center of mass and they had dropped like stones. Capt. Jaxx counted the shots as he always did even when firing on full automatic—trained soldiers do that. When ammunition is in limited supply and may be difficult to replace, round count and trigger control is important.

The good news for Capt. Jaxx was that so few rounds of steel-jacketed 9mm ammunition did not do a lot of damage. The Germans' multigreen summer pattern camouflage smocks, reversible to winter white snow, were in fairly good shape. The bad news was both had blood on them even though it had dried in the sun.

Capt. Jaxx studied the three-quarter-length smocks and selected the least bloody one and stripped it off the dead Gebirgsjäger. He put it on over his faded green jungle jacket then draped all the bandoliers of ammunition he had acquired over the dark spots as best he could. Since the smock had been worn summer camouflage side out, the dark reddish brown bloodstains were not overly noticeable. He hung the 6x30 Zeiss glasses around his neck. Then he put on one of the dead men's billed caps with the skier's round sun goggles strapped to the front of the flat-topped crown and moved out down the trail toward the village. If it were not for the blue jeans bloused into his canvas-topped raiding boots, he could have easily stood formation as a 1st Mountain Division trooper—as long as he didn't say anything.

The trail down from the church was treacherous, steep and winding. But it was a lot easier than not using it in that rough terrain. When he reached the village the few old men and women in the streets drifted away. The 1st Mountain Division had the unenviable record of having executed more prisoners and civilians in occupied lands than any other Wehrmacht unit in the Aegean Theatre of Operations. Even more than the Criminals of the 999th Light Division. Understandably the villagers tried to limit contact with them.

Capt. Jaxx's attempt to masquerade as a Gebirgsjäger seemed to be working.

4

BOREHOLING

MAJOR THE LADY JANE SEABORN SAID, "I UNDERSTAND you no longer intend to fly to No. 4 Middle East Training School to lead the OSS lads back here to make their first jump. An idea, I might add, that ranks as one of your all-time worst ever. You know those troops have not received adequate training."

Colonel John Randal looked up from where he was cleaning his pistols on the coffee table in the living room of their second-story suite in ABCHQ.

"General McKoy vetoed the idea of me going."

Which was not exactly true.

Lady Jane said, "I realize rescuing Colonel Roosevelt's North Africa Photo Reconnaissance Wing pilot has high priority, but is it absolutely necessary for the OSS lads to perform the mission?"

Col. Randal said, "It is unless you can find me another team of operators who can arrive on ABC in the next few hours armed, equipped and ready to fight, Jane."

Lady Jane said, "In that case, Beverly and I are left with no choice but to have one of the duty pilots fly us to Kabrit. We shall link up with Det 3, travel straight back here and jump in with the lads to keep their spirits up. Someone senior from our headquarters needs to show the flag.

"I shall expect you to be on the DZ to retrieve my parachute and carry it to the assembly point—Beverly's as well."

Col. Randal said, "Whoa . . ."

But Lady Jane was gone.

What just happened?

Not for the first time the thought occurred to him that commanding the women of Raiding Forces required a skill set radically different from anything he had been trained for. Col. Randal was not sure he had ever actually commanded any of them.

CAPTAIN BILLY JACK JAXX WALKED DOWN THE CENTER OF the single street in the village. If there were any people around, they were staying out of sight. It was not any different from the other islands he had been on. Life for civilians on German-occupied islands in the Aegean was a nightmare for the civilians. The criminals of the 999th Light Division were crazy—literally. Some had been released from mental institutions for the criminally insane if they were determined to be capable of performing basic military duties. Out of sheer boredom they had been known to occasionally shoot people at random for fun

The 1st Mountain Division Jägers were worse. Hardened combat veterans, they were brutal occupiers who did not waste their time killing people in ones or twos. Any sign of resistance, real or imagined, and they would line up fifteen or twenty locals and shoot them in retaliation. The division had a cold-blooded record of executing Greeks and Italians *en masse* with little or no provocation and a complete absence of remorse. No wonder there were not a lot of locals out and about.

There were not any Nazis in sight either. Where might they be? Probably laying low on the theory, "if they can't see you they can't hit you." Meaning being put on a work detail designed to make the NCO who cooked up the meaningless project feel like he was doing his job.

Some soldiers believe the concept of "if they can't see you they can't hit you" also applies to combat. Capt. Jaxx knew that theory was not true. There were a lot of ways to get tagged in a combat zone even when out of sight—like on a submarine.

He came to the building he had identified as the 1st Mountain Division Detachment Command Post. No activity was immediately visible. That was fine with him. He knew Germans were in there from having observed them entering and keeping the building in line of sight on the way down the hill and not observing anyone leaving. Germans were still inside.

Capt. Jaxx turned right and walked toward the two-story building directly across the street. He went around back and climbed up on the roof. It was fairly easy due to the steep slope of the escarpment. All he had to do was step off the ground onto the roof. They were almost touching each other. Once on top he scanned the village but did not see any 1st Mountain Division personnel anywhere. He did see a few Greek men moving around off in the distance staying well away from the central section of the town.

Capt. Jaxx unslung the Russian Tokarev SVT-38 self-loading rifle, placed the weapon against the decorative four-foot-high parapet running around the flat-topped roof on the street side and dropped the bandolier of 7.62 ammunition next to it. King had briefed him on the weapons' capabilities after Raiding Forces' patrols captured a few of the Tokarev rifles from Afrika Korps units with prior service in Russia. Some German outfits had appropriated them for their own use while serving on the Russian front.

The Merc had encountered STV-38's during the Winter War in Finland when he was fighting the Russians. He pointed out the weapon had several design flaws, not the least of which was the magazine fell out occasionally. With a Russian-manufactured 3.5x21 PU scope mounted, the firearm was intended to be utilized as a precision sniper rifle. However, after only a few rounds it developed lateral dispersion, stringing its shot pattern. A no-go for the long-range sniping out to 1,300 yards as envisioned by the Red Army.

According to King, when Hitler ordered the invasion of Russia, the Germans captured Tokarev 7.62 SVT-38s by the hundreds of thousands, by some estimates up to a million. Field expedient modifications by unit armorers could cure most of the weapons deficiencies but not all of them. Nevertheless, the Wehrmacht was constantly in need of small arms of all makes and models, so the SVT-38s were pressed into service.

The self-loader worked reasonably well as an infantry service rifle as long as it was kept to ranges under 600 yards and you kept an eye on the magazine. None of that mattered to Capt. Jaxx.

He was not a long-range sniper.

VERONICA PAIGE ARRIVED AT THE SUITE. FLANIGAN showed her in. Colonel John Randal's instructions to the former Habbaniya police officer were to allow in anyone who was not a Brandenburger Sea Raider. The ex-policeman doubted the colonel actually meant that, so he took the precaution of first calling on the phone to announce an incoming visitor.

Veronica said, "You wanted to see me, Colonel?"

Col. Randal said, "How many times am I going to have to ask you to call me by my first name before you get the idea I might actually like it if you did that?"

Veronica said, "As a military wife, I respect your rank. And Colonel, I have come to . . ."

Col. Randal said, "You do understand in the best British regiments the officers address each other by their first name regardless of rank unless they're 'in front of troops.' And that means in the act of commanding a formation, a parade or leading in action."

"I am aware of the tradition," Veronica said. "However, the way you have mentored my daughter has earned my gratitude. I salute *you,* not your rank. From the first day Mandy met you at Habbaniya she blossomed into this extraordinary young woman I almost do not recognize as my own."

Col. Randal said, "From now on you call me John unless I'm in front of troops. That's an order. Feel free to quote me on it. I'll put it in writing if that's what it takes."

Veronica laughed. "Surely my calling you by your first name is not what you requested to see me about."

"Roger that, it's not. General McKoy explained to me the political ramifications of rescuing the NAPRW pilot downed on Sikinos.

Apparently, the President and the Prime Minister both have an interest in the outcome as well as General Donovan, the Chief of the OSS. Typically a mission of this type would fall to MI-9. You would supervise the collating of intelligence on the target and conduct the initial workup on the situation, then hand it off to Small Raids Inc. for the logistical workup and Raiding Forces for the tactical planning. Then I would take over responsibility for the execution of the mission.

"This time's going to be different so I wanted to give you a heads-up —avoid any misunderstandings."

Veronica said, "You gave me the Escape assignment at a time no woman had ever done anything like it before. Then backed me to the hilt when London attempted to replace me with one of their male MI-9 officers. There is no inter-service rivalry between us over bureaucratic territorial boundaries, John. And there never will be. I work for you no matter what any TO&E chart says."

"I'm not taking your mission away. I'll just be more involved than usual on this one from the start," Col. Randal said. "You'll be in on all the planning and mission prep right up to LD time. Once this operation is over, if successful, my after-action report will reflect it was an MI-9 Escape extraction of a downed Allied pilot from enemy-controlled territory supported by Small Raids Inc. with the ground maneuver element provided by Strategic Raiding Forces."

"And if it fails?"

"I'll try not to write a report."

COLONEL JOHN RANDAL PLACED A CALL TO VICE ADMIRAL Sir Randolph "Razor" Ransom's new WREN, Second Officer Blyth Colthorp, the daughter of one of the Seaborn families longtime friends. "Would you inform the Admiral I'd like to come see him at his earliest convenience?"

"Wait one . . . sir, Admiral Ransom will see you now."

"Tell him I'm on the way down, Second Officer," Col. Randal said.

Outside the door to his suite Col. Randal found Lieutenant Chase Starrett talking to Flanigan at the security desk.

"What are you doing here, Lieutenant?"

"Waiting for you, sir."

"What can I do for you?"

"Sir, you told me I was your temporary aide so I'm standing by. . ."

Col. Randal said, "Walk with me."

The two started down the stairs.

Col. Randal said, "So here's what I want from you—do what Capt. Jaxx does. He keeps dialed in on everything that's going on. If I plan to leave the building I always stop by the TOC and check in with him. He briefs me on any new developments or anything else he believes I need to know. Jack's got a sixth sense for identifying problems *before* they crop up. When there's something he believes I need to know he comes and finds me.

"I want you to be my eyes and ears but you're not my spy. Tell me what you observe and anything else you feel I should be aware of. *Never,* ever tell me what you believe I want to hear. And I don't need a flunky to shine my boots—clear?"

Lt. Starrett said, "Good to know, sir. Your field boots look like the last time they got any polish on them was the day they came from the bootmakers. They're almost snow white."

Col. Randal said, "Yeah, well, we don't want moonlight reflecting off the polish."

Lt. Starrett said, "That's what the instructors told us in Achnacarry when I was at Commando School—you think that could really happen, sir?"

"You ever see moonlight reflect off of anything other than a lake, Lieutenant?"

"Negative, sir."

"Well, neither have I."

"Why take a chance, sir?"

"You're picking up on the finer points of this aide business pretty fast."

Downstairs, WREN 2/O Colthorp said, "Go right in, Colonel."

VAdm. Ransom was engaged in a conversation with Lieutenant Ted Hamilton, OBE, aka "The Great Teddy." The two were discussing the merits of a SECRET paper recently circulated by the Department of the U.S. Navy titled *Gambit Tactics*. It outlined a simple but possibly effective technique for antisubmarine patrol aircraft to lure an enemy submarine back to the surface after a U-boat the plane spotted had crash-dived.

VAdm. Ransom wanted Lt. Hamilton to review the paper to see if he could suggest any refinements. Gambit Tactics were essentially a sleight of hand. Something The Great Teddy was a genius at performing.

Col. Randal said, "Could the Admiral and I have the room?"

It was not really a question. Lt. Hamilton and Lt. Starrett made themselves scarce. They went outside to wait in the lobby of the TOC, which gave them an opportunity to flirt with 2/O Colthorp.

VAdm. Ransom said, "You requested to see me, Colonel?"

Col. Randal said, "Yes, sir . . . this conversation never happened."

VAdm. Ransom showed no outward sign but his eyes clicked. "Aye, aye…understood."

Col. Randal said, "You'll be the closest thing to a father-in-law I'll have once Jane and I are married, Admiral, so I'd like some father-in-lawyerly advice prior to."

VAdm. Ransom appeared taken off guard . . . rare for the Razor. "I had no idea you would feel that way, Colonel. Full steam ahead. I shall do my best though I freely admit I was a complete failure as a parent. Do not expect much."

"Are you familiar with the term 'boreholing,' sir?"

"Not that I can recall offhand."

"It's a form of punishment in a POW camp meted out by the inmates to a fellow prisoner who commits an act considered to be a capital offense. The guilty party is held upside down, stuffed down a latrine headfirst and drowned."

"What type of offense?"

"Stealing food in this case."

"That would justify the sentence. Worst crime a man can commit in a POW camp. Why are you asking?"

"Jim informed me intelligence has confirmed Jane's husband Mallory made it ashore after his ship was sunk. He was imprisoned in a Japanese..."

VAdm. Ransom interrupted. "Got caught stealing food. Man always was a bloody fool. Never understood what Jane's father saw in him."

Col. Randal said, "We decided against informing Jane on the grounds the means and methods used to acquire this information is highly classified. That said, it seems like a flimsy excuse for not advising a woman she's a widow. I try not to keep secrets from her, but there's no reason I can think of to disclose this information right now—good excuse or not."

VAdm. Ransom said, "I concur."

"I'd like your recommendation on how to handle this, sir."

VAdm. Ransom said, "We could always say Mallory died in the camp. The problem is the other prisoners are never going to confirm the story. After the war they shall all claim he simply disappeared."

"That's what Jim says."

"Glad you brought this to me, Colonel—family business. I like that. You and Jim made the right decision," VAdm. Ransom said. "This is the type of incident that needs time to play itself out. We do not tell Jane until a more appropriate moment presents itself—as in never."

Col. Randal said, "Thanks, Admiral."

VAdm. Ransom said, "When you and Jane are eventually married the two of us are going to get along famously. I liked the cut of your jib the first time we met. Told you so, as you may recall."

"Yes sir, you did."

He knew the Admiral had never liked the cut of anyone's jib on first meeting in his entire life. The Razor had merely encouraged him to feel free to pursue a relationship with Lady Jane despite the fact she was a titled member of the British aristocracy—something he did not have to do. The two of them had gotten off to a good start.

As Col. Randal was leaving, VAdm. Ransom decided not to mention the *Perseus* had missed a second *and* third radio check. Failing to inform him was a calculated judgment based solely on exigent circumstances. The submarine was now officially declared "missing." By standing tradition in both the United States Navy and the Royal Navy subs are never declared "lost" unless wreckage was found—only "missing." The Razor made an on-the-spot command decision that it was in the best interest of Col. Randal—meaning the best interest of Small Raids Inc./Raiding Forces—not to be distracted by finding out Captain Billy Jack Jaxx would not be coming back from his cruise.

The mission to recover Colonel Elliott Roosevelt's photo reconnaissance pilot took precedence over everything. The implications of a successful rescue operation for the future of Small Raids Inc./Raiding Forces were too far-reaching to be compromised by the tidal wave of emotion that was going to come crashing down on the men and women of both units once the news of Capt. Jaxx's death broke. Nothing could be done about it—there would be time to grieve later. VAdm. Ransom needed Col. Randal to be free of distraction to give the mission his full attention.

He also decided now was not the time to ask the one burning question about the immanent recovery attempt: Why was the Colonel bringing in OSS OG Det 3 to conduct the operation? VAdm. Ransom could not rationalize this move. The Frogs were the least experienced personnel in Raiding Forces. Surely Col. Randal could have cobbled together a team from the veteran troops stationed aboard the Levant Schooner Flotilla (LSF). The Razor refrained from asking but it was not easy—restraint not being a quality he was known for.

One thing he was certain . . . Col. Randal had a reason.

MAJOR "PYRO" PERCY STIRLING, DSO, MC, HAD JOINED THE conversation at Second Officer Blyth Colthorp's desk when Colonel John Randal left Vice Admiral Sir Randolph "Razor" Ransom's office.

"The Admiral is ready for you again, Lieutenant Hamilton."

"Sir!"

"Let's step over to a corner, Major."

Lieutenant Chase Starrett wondered if this conversation should include him. After a split second's hesitation he decided it did. It was the right call.

Col. Randal said, “Terry is being recalled to England with the Lancelot Lancers to prepare for the invasion. Stephanie has recommended you as his replacement. Lady Jane, Admiral Ransom and General McKoy have concurred. What do you say?”

Lt. Starrett was shocked to hear the nominating process included two women—nothing against them. The U.S. Army was essentially an all-male organization. While there was the Women’s Auxiliary Corps (WAC), no commander in his right mind would have consulted a WAC on who to appoint as his Operations Officer. Or if he did he would never admit it. Certainly not to the man selected for the job.

Maj. Stirling said, “I have a rather hard time imagining Raiding Forces without Zorro, sir.”

“So do I.”

“I shall accept the assignment bang on with regrets, Colonel.”

“Not the best choice of words, Pyro.”

“Quite right, sir. Strike the ‘bang.’”

“Go see Terry. Then get with Stephanie. She’ll be expecting you.”

After Maj. Stirling walked off, Col. Randal said, “You have a question, Lieutenant?”

“Yes, sir. Is Major Stirling going to be all right with taking a job assignment endorsed by a woman junior to him? I have a hard time picturing something like that in the U.S. Army, sir.”

Col. Randal said, “Without Lady Jane’s social connections and her Royal Marines to perform the jobs we don’t have the manpower for Raiding Forces wouldn’t exist. Capt. Fawcett-Tatum is almost as well-connected and not the least reluctant to use her station in British society for our benefit. Major Stirling knows that. If he has a problem with who signed off on his new job he’ll get over it.”

Col. Randal did not mention he offered the S-3 Operations assignment to Captain Stephanie Fawcett-Tatum first—he had no problem with her being his operations officer. That information was above the lieutenant’s pay grade. Personally, he wished she had taken the job. Major Stirling was a brilliant squadron CO. He was needed to command it.

Now Col. Randal was going to have to find his replacement.

LIEUTENANT JACKSON TAYLOR, THE HOLLYWOOD dentist, had been called to London. He was a fluent French speaker. The OSS had an undisclosed assignment for him somewhere on the continent of Europe. Because the Secret Intelligence Service (SIS) and SOE considered the European Theatre of Operations (ETO) their private preserve and had been freezing out the OSS, the opportunity to put one of his agents on the Continent was too tempting for Brigadier General William "Wild Bill" Donovan to pass up.

That left Ensign Westly Slade as the senior officer—in fact the only officer—in the Office of Strategic Services Maritime Unit Operational Group Detachment 3. A lieutenant from the super-secret Royal Marine Boom Patrol Detachment was being reassigned to Raiding Forces at Vice Admiral Sir Randolph "Razor" Ransom's request. The initial idea was to employ him with Major Butch Hoolihan's Red Indians. Now the thinking was he might be best assigned to OSS OG Det 3. Time would tell as the situation developed.

Since Colonel John Randal had a policy that officers lead, Ens. Slade was currently at No. 4 Middle East Training School (No. 4 METS) located on Kabrit Royal Air Force Base, which was shared with the United States Army 9th Air Force—leading, though not exactly by example.

Ens. Slade was standing sideways on a ten-foot platform inside an open-air metal canopy pavilion constructed over a giant sandpit. The Frog officer was strapped in a parachute harness with no parachute pack or reserve. The risers running over his head were attached to a large wooden bar that looked suspiciously like a trapeze. The "Swing Landing Trainer" had been designed by an Airborne sadist whose name had been lost to history. The apparatus' purpose was to teach the gentle art of the Parachute Landing Fall.

There were four basic ways to exit the platform correctly. From the front, from the right or left side and from the rear. This being the first full day of a drastically shortened Jump School program, Ens. Slade was making a right-side exit off the platform. The idea was for him to perform a "straight right side" Parachute Landing Fall or PLF as it is affectionately referred to by paratroopers. A right-side exit off the platform is the easiest to make unless the jumper is left-handed. For that reason it was a good place to start the novice paratroopers's training. However, the probability of actually making a straight on right side PLF—meaning coming in sideways, after an actual parachute jump—was in the range of approximately zero.

Having no real control over their parachutes other than to slip right or left, on most jumps parachutists came in backward, something not widely advertised in Jump School, with coming down forward being a distant second. Performing a right or left side PLF requires a jumper coming in backward or forward to land on the balls of their feet and, on hitting the first "Point of Contact," swivel to the right or left—which requires an instantaneous decision by the jumper —deciding which type of PLF, right or left, immediately prior to slamming to the ground. Then, decision made, the jumper is expected to swivel and fall "as limp as a dishrag," hitting the four remaining Points of Contact, which are located on the side of the calf, the side of the thigh, the side of the buttocks and finally on the meaty portion of the back, all the while remembering to "keep those elbows in." The head is not one of the approved Five Points of Contact.

A PLF resulting in anything other than that exact sequence is a flunk—meaning you may break at least one major bone or end up knocked unconscious, neither of which is a desirable result when parachuting behind enemy lines or making a forced entry airborne assault in the immediate vicinity of an enemy position. If a paratrooper is coming in for a landing on the right or left side it means he is likely in a hurricane and everything he has been taught up to that point is off the table.

Executing a successful PLF is a lot to process . . . and as was said in Raiding Forces a lot, "Not for the faint of heart." Preparatory to landing the jumper has to decide on the best course of action, then execute it in an incredibly compressed window of time, traveling at speed toward disaster while wondering about what the enemy might be doing simultaneous to his landing. All this is especially tricky for anyone afraid of heights or who had been airsick on the flight to the DZ and was dealing with that. Any person who did not have reservations about jumping out of a perfectly good airplane was suffering from acute mental or emotional issues worth noting. Even though he was only ten feet up it was perfectly understandable for Ens. Slade to have concerns.

In fact he was sweating bullets.

At this point in the OSS OG Det 3 student's jump training evolution the Frogs had not learned what steps to take in the event of a malfunction such as a Mae West, a Roman Candle or an inverted canopy, or how to spill air out of the canopy to negate excessive wind drift or how to deploy the reserve chute in the event the main malfunctions so that it does not simply fall out, drop down and wrap around the jumper's ankles as immortalized in the U.S. Army Airborne's celebrated song *Blood On the Risers* with its cheerful chorus of "gory, gory what a hell of a way to die." The OSS operators would get to those later.

From the instant a paratrooper exits the aircraft he begins counting to himself, "one thousand, two thousand, three thousand,"—on reaching "four thousand" he immediately begins making his safety checks. The first being to "check canopy," which is of primary importance. The parachutist has to recognize there is a malfunction, decide what type he is experiencing and what countermeasure to take, then execute the school-solution corrective action before slamming into the ground at terminal speed, not forgetting to hit his Five Points of Contact, while keeping his feet and knees together but *not* locked, head down, chin on his chest, elbows in—improvisation not being recommended.

Becoming a paratrooper is not for slow thinkers. Brain freeze due to duress is a drawback. Every jump is different so do not expect them to ever become routine.

The Swing Landing Trainer was attached to a rope held by an instructor. On being given the order to "Go"—the last command a jumper hears before actually exiting an aircraft—the trainee was expected to exit the platform in the approved body position: head down, chin on chest, feet and knees together slightly bent but *NOT* locked, with hands on the ends of the reserve parachute on his chest, elbows in. Since he was not wearing a reserve, the hand on the reserve part was simulated. The Airborne Instructor would let him swing back and forth a few times then announce in a loud vigorous voice what type of PLF the student should execute, give the command "Prepare to land—land" and let go of the rope—BOOM!

This being the first attempt of the morning and because he was being used as a demonstrator for the rest of the class, Ens. Slade was informed "RIGHT FRONT PLF" prior to jumping off the stand so he would know what was coming. For the rest of the training cycle the student would have no inkling of which landing fall he is supposed to attempt until *after* he exited the platform and was swinging back and forth immediately prior to the instructor releasing the rope. Again, the options being: right front, left front, right rear or left rear.

There is no such thing as a straight front or straight rear PLF in the U.S. or British Airborne even though the 1941 Hollywood movie *Parachute Battalion* showed clips of actual pioneering Airborne students doing forward somersaults in the very early days of jump training—they were still trying to figure out how to do it at Ft. Benning.

In Germany Fallschirmjägers were taught to do a forward roll upon landing because they only have a single riser attached in the back of their harness and the jumper dangled down face first, hands nearly touching the toes of his boots during the ride to the ground. On anything other than a manicured lawn the forward headfirst roll PLF is basically suicidal if not performed exactly right. Captain Mike "Mad Dog" Reupart, that legendary trainer of military personnel who desire or maybe not to become paratroopers—the Frogs having never been accorded the customary option of volunteering—advised, "Do not try the German method at home."

Roger that.

On the Swing Landing Trainer, when the instructor releases the rope, what invariably happens is the student crashes to the ground, usually ending up in a crumpled pile, having dinged up parts of the body he or she never knew existed. It takes repetition, repetition, repetition, followed by quite a lot of additional repetition, repetition, repetition, to develop the requisite muscle memory, most of which is recall of pain, to become proficient at parachute landing falls. Some paratroopers go their entire career without mastering the technique of how to perform anything vaguely resembling an acceptable PLF. Especially those performed backwards starting five or ten feet in the air off a Swing Landing Trainer with your body parallel to the ground when the rope is released by your heartless instructor.

The major drawback to the "repetition, repetition, repetition" method of developing muscle memory as it pertains to executing a PLF is that the repetition, repetition, repetition hurts. The student jumper is layering pain on top of pain. The upside is, once mastered, the skill is never forgotten. It becomes automatic.

Capt. Reupart gave the peremptory command "Go!" Ens. Slade hopped off the platform in the correct body position simulating as much as possible a vigorous exit from the door of an aircraft only this was inhibited by the inconvenient fact of standing sideways. He was in a reasonably good tuck position for a student in his second half-day of training. When, with his knees bent and the toes of his boots swung up pointing straight toward the roof of the building, he arrived at the apex of his swing, Mad Dog shouted, "PREPARE TO LAND, RIGHT *REAR* PLF, LAND!" and let go of the rope. He had changed things up to see if Ens. Slade was paying attention.

Not that it mattered.

Major the Lady Jane Seaborn and Beverly Blackwell arrived at the facility in time to witness what might qualify as the worst Right Rear PLF in modern Airborne history. Ens. Slade managed to not hit a single one of the Five Points of Contact in any sequence. Which is pretty much impossible. He was lying flat on his back having a difficult time breathing, wondering how his military life had come to this.

While still in college at Dartmouth, Ensign Slade was recruited by OSS because of his aquatic skills. After two years of intensive training he was supposed to be somewhere in the Pacific swimming off a Japanese-occupied island marking landing sites, clearing lanes and blowing up obstacles in advance of Marines storming the beach. Not in the desert learning how to jump out of airplanes at a joint RAF/U.S. airbase near a town called Kabrit which when translated into English meant sulfur.

Hell was said to smell like sulfur.

Ens. Slade did not realize he was having the best part of his day.

CAPTAIN BILLY JACK JAXX CLIMBED OFF THE BUILDING leaving the Russian 7.62 SVT-38 self-loading rifle leaning against the parapet on the roof along with one of the MP-34s. He walked around to the front and across the street to the 1st Mountain Division Detachment Command Post. Having carried a 9mm Steyr MP-34 submachine gun all day and using it in the prone position from time to time as he conducted observation of certain locations of interest, he was beginning to change his opinion of the weapon somewhat. Even though awkward the side-mounted magazine had its advantages.

While not entirely sold on the weapon he was beginning to warm up to the 9mm Steyr MP-34. Unlike his personal submachine gun of choice the 9mm Beretta M-38, or even the .30 M1 Carbine he was carrying more and more often these days, the Steyr allowed him to lay perfectly flat due to there not being a magazine protruding from the magazine well on the bottom. An important consideration when being shot at.

And it allowed him to put out grazing fire only inches off the ground.

He had one of the MP-34s slung over his shoulder and another in his hands. There was no one outside the CP, but then he already knew that. Capt. Jaxx had only taken his eyes off the entrance for a few seconds as he came around from behind the building. Not long enough for a German Jäger to exit without being seen.

Capt. Jaxx had enough of sneaking and peeking. Hiding and watching from concealment for extended periods was not his idea of a good time. He walked up to the open front door of the CP having already unscrewed the metal cap at the base of one of the wooden-handled Model 24 grenades he had captured. Pulling the loop on the braided cord concealed underneath until it snapped, he unceremoniously tossed the grenade inside the building. The German Model 24 had a four- *or* five-second fuse—it varied from grenade to grenade. With no identifying marks to indicate fuse length, there was no way to know which one. And since German munitions were assembled by slave labor, they *might* have been cut shorter accidentally on purpose as a nasty surprise for the thrower.

It was not a good idea to pull the fuse lighter and "cook one off" in your hand prior to tossing—even if that technique was all the rage for grenade throwing GIs in the patriotic Office of War Information (OWI) approved war movies coming out of Hollywood these days. Still, after he had thrown in the first one, he had enough time to unscrew the cap on a second grenade, uncoil the detonating cord, jerk the pull loop, pitch the grenade inside and step out of the way of the open door to avoid the first potato masher's back blast.

The building shook when the initial grenade detonated. Smoke and dust blew out the door and open windows. The German Model 24 was noted for having more blast than shrapnel effect. In the enclosed space of the CP, it made a heavy-duty, though highly muffled, *WHOOOMPH!*

Followed by a second *WHOOOMPH!*

Capt. Jaxx stepped in through a cloud of smoke and dust. He spotted four Nazis in various stages of distress. Two of them on the floor might have been dead. Of the remaining two, one was lying across a desk moaning and the other was at another desk slumped in a swivel chair trying to raise his hands to his bloody face.

Capt. Jaxx immediately commenced fire. The first to receive a burst was the Gebirgsjäger in the chair. He hosed him with 9mm rounds remembering, as always, to count how many were fired—twelve—center of mass. That might have been excessive. He shifted to the Nazi lying face down on the desk, gave him two bursts of three in the back and then emptied the rest of the magazine into the two on the floor. Six rounds each. Trigger control and knowing your round count is important in a firefight even when the other side is not shooting back.

Instead of reloading, he slung the Steyr M-34 over his left shoulder on its carrying strap while simultaneously unslinging the one on his right. In a close-range dangerous encounter there is no such thing as being too ready. Not to worry—everyone in the building was now KIA. Or as Capt. Jaxx liked to say, "seriously dead."

With the fully charged submachine gun (SMG) at the ready, he went back outside, walked across the street and climbed up on the roof of the building again. Only then did he reload the empty 9mm Steyr M-34 submachine gun.

Nothing happened for a few minutes. Then two 1st Mountain Division troops came jogging up the street toward the CP. They had heard the muffled grenades and SMG rounds but being in the distance, inside a building, and with the breeze blowing off the Aegean distorting the sounds, the Nazis were not entirely sure what they had heard.

Despite the Germans being veterans of half a dozen campaigns they were not overly alarmed. That was apparent by the casual way they carried their bolt action 7.92 Karabiner 98K rifles. Capt. Jaxx believed they most likely did not suspect the Greeks had somehow obtained weapons and attacked the CP. The locals knew what would happen next should something like that ever take place. The Jägers most likely believed there had been an accident. Those are possible with grenades stuck in your belt or jackboots or sometimes when bored troops are fooling around with their demo and something goes wrong. It happens in the best of units.

Crouching down from his position behind the parapet, Capt. Jaxx watched the two Nazis approaching. When the men were almost directly below, he leaned over the top of the parapet with the 7.62 SVT-38 self-loading rifle and shot them both. *BLAAAAAM! BLAAAAAM!*

The gunshots reverberated in the confines of the narrow cobblestone street with zero lot lines for the buildings facing it built like duplexes with no setback—loud. No one was going to mistake them for anything other than what they were. The Nazis dropped in their tracks. Their heads appeared to hit the ground before their feet left it. You do not see that in the OWI movies. It did not happen in the Saturday matinee cowboy movies he loved so much growing up either.

Seeing someone shot by a high-powered rifle always seemed unreal to Capt. Jaxx even when he was doing the shooting. The breeze ruffled the collar of one of the dead Jäger's reversible camouflage smocks so that the snow camo side flashed white from time to time. It was the only movement he could detect anywhere in sight.

The village was very still.

AS ENSIGN WESTLY SLADE STRUGGLED TO RECOVER FROM his run-in with the swing Landing Trainer the Operational Swimmers listened in disbelief to what drop-dead gorgeous Major the Lady Jane

Seaborn had to say following their being instructed to fall in around her. The Frogs were not the kind of men to admit to being horrified—but they were now.

Lady Jane said, "Red Alert."

Raiding Forces had three types of alerts: Stand By, Stand By Ready, and Red. The unit was on Stand By Alert around the clock. Day and night seven days a week—it meant always being ready for a mission. Stand By Ready Alert was an escalation in preparedness that meant those so alerted were going on a mission, only waiting for the movement order to saddle up and go. Red Alert meant move to the point of debarkation for transport, mules, jeeps, boats, planes, etc.—*now*!

The Frogs were aware of the different phases of Raiding Forces alerts. In the short time they had been assigned, all of the OSS operators had gone on at least two missions. Ens. Slade had made three. Since Raiding Forces was what was called a "high-speed low drag" outfit, the alert sounded like the real thing . . . only it *had* to be part of the training exercise.

What Lady Jane said was like getting zapped by a bolt of lightning or being strapped in an electric chair when the warden threw the switch. "Det 3 is to immediately move to a departure airfield where you lads shall board a Troop Transport Command C-47 and be flown to Castelrozzo. Once over the island you are to drop by parachute and reassemble on the DZ.

"After assembling you are to move overland to the dock at ABCHQ. Once there you shall board our amphibious Hudson aircraft for a flight to a rendezvous with an RAF Air-Sea Rescue High-Speed Launch at sea. Upon linking up with the HSL you offload, transfer to the crash boat for transportation to German-occupied Sikinos Island.

"Upon arriving Det 3 is to execute a forced entry amphibious assault landing ashore to retrieve a downed South African Air Force Photo Reconnaissance pilot known to be at large on the island.

"Questions?"

The Frogs were so traumatized by the thought of parachuting onto ABC no one could think of any of the long list of questions they should have asked about the mission.

"No? In that case, lads, be not daunted when you fail to remember everything you have been taught about parachuting. You most definitely shall. Not to worry, it is almost impossible to make a proper PLF without eons of practice. No such thing, actually. You shall be fine.

"A crew standing by on the DZ is there to help recover your chutes in the event you have trouble with the Quick Release System due to the condensed nature of your training."

Training?

"It may be of interest to know the U.S. President and the Prime Minister of the U.K. are both monitoring the operation. Colonel Randal specifically selected Det 3 for this extraordinarily high-profile assignment. Do not let him down nor myself or Beverly . . . your biggest champion, by the way."

Beverly said, "Lady Jane and I are jumping on ABC with you—it'll be fun."

Was she crazy?

Captain Mike "Mad Dog" Reupart ordered, "Repair to your quarters double quick, secure arms and equipment, report back here in five minutes to be trucked to the departure airfield—MOVE OUT!"

This was no drill.

5

THERE WENT PLAN B

VICE ADMIRAL SIR RANDOLPH "RAZOR" RANSOM WAS standing in front of the large ATO map studying Sikinos Island with James "Baldie" Taylor and Doctor Layton Winthrop. Of interest was the small unpopulated spit two miles offshore where LRDG Beach Watch Baker Team 6 was located. Colonel John Randal walked over to join them. Lieutenant General "Geronimo" Joe McKoy noticed the gathering and interrupted the conversation he was having with Captain Stephanie Fawcett-Tatum to drift over to see what was taking place. Veronica Paige followed.

Dr. Winthrop said, "I confirmed with the Long Range Desert Group that Baker Team 6 consists of a six-man Beach Watch party. They have the capability of transporting me via rubber assault raft the two miles to Sikinos to link up with my agent. Unfortunately there is no way to notify him in advance of my arrival.

Col. Randal said, "With Pam traveling and Beverly committed to delivering Det 3 to the HSL, do we have a pilot available to fly Dr. Winthrop to the Baker Team 6 location?"

Captain Stephanie Fawcett-Tatum, who had joined the group, said, "I sent for Wing Commander Wilcox. He shall be arriving later this morning. The remainder of our Special Duties pilots are currently away from ABC on missions stretching from Cairo to Greece."

Lt. Gen. McKoy said, "I've discussed the situation with the Admiral. At a bare minimum we need the Hudson and a Walrus to pull this off. Lose either of 'em for any reason—mechanical problems, pilot gets sick—this mission has to stand down. At least until we can get some kind a' replacement in here."

V.Adm. Ransom said, "With all our current commitments we are stretched perilously thin. We have never been so overtaxed with no relief in sight. Additional amphibious troop-carrying aircraft are not able to arrive until sometime tomorrow. We need to go tonight, people."

Col. Randal said, "No way to pull this off using HSLs or our MAS boats?"

V.Adm. Ransom said, "Negative—not unless we are prepared to operate in broad daylight."

There went Plan B.

OFFICE OF STRATEGIC SERVICES OPERATIONAL GROUP Detachment 3 loaded onboard a Troop Transport Command (TTC) C-47 for the flight to Castelrozzo Island. Major General Sam Houston "Bronc" Blackwell had made arrangements for TTC to supply troop carriers to any airborne unit or school that required planes for parachute training or refresher jumps. Unlike previous TTC commanding officers at all levels who resented the training commitment as a drawdown on their assets, Maj. Gen. Blackwell saw them as an opportunity. He made his thoughts clear to his subordinate commanders. Bronc wanted his pilots and aircrew to have the benefit of as much practice as possible dropping parachutists in order to be ready when the big day—most likely a night—came to invade the continent of Europe.

Maj. Gen. Blackwell was a CO known to and beloved by his troops. They openly called him "Bronc Blackwell," all said as one word or he was called "The Bronc." A real-life living legend in the USAAF—the Commander of Troop Transport Command was always described as "the *legendary* Bronc Blackwell" when his name appeared in the press—as it did more and more frequently nowadays. Not some remote godlike figure as were most general officers, Bronc was a hands-on, lead-from-the-front commanding officer.

Captain Richard Holden, the Dakota's pilot, and his co-pilot, Lieutenant Malcolm North, were cognizant of the fact that Bronc's daughter was onboard today and would be making the jump. The two USAAF officers were more than slightly traumatized by the prospect. The word was, and it was widely understood throughout TTC, the General had a good-looking girl but no one had said she was *that* good-looking. When D-Day finally arrived, a highly contested night combat parachute drop on the continent of Europe was going to be a less stressful mission for the two pilots than being responsible for flying to a tiny flyspeck of an island off the coast of Turkey and putting Bronc Blackwell's beauty queen daughter out the door on time and on target. This flight may have been billed as a training exercise but things can and do go wrong.

In the troop compartment, the Frogmen were sitting on nylon bench seats running the length of both sides of the airplane. They had a 250-mile flight ahead of them. The Operational Swimmers were not wearing their parachutes—which they had yet to even put on at this abbreviated stage of their Airborne training. Today it would be onboard rigging which is a complicated affair even for experienced paratroopers. Captain Roy "Mad Dog" Reupart and two of his staff were on the plane to help the Frogs chute up and to give each of them a jumpmaster inspection. Capt. Reupart would be acting as the jumpmaster today. He and his assistant jumpmasters were static, meaning they would not be exiting the aircraft. To say tension in the troop compartment was thick enough to "cut with a knife" would be no exaggeration. A much-abused description to depict high-stress situations, this was one of those times it actually was.

Before taking her seat at the rear of the C-47 as it was starting to taxi, Major the Lady Jane Seaborn flashed one of her signature heart-attack smiles and announced, "It is tradition for U.S. Paratroopers to lock arms and hold their boots off the deck as the plane takes off. The ritual originated with the glider units. The idea being to assist the unpowered glider lift off."

That sounded totally insane to the Frogs but not more so than anything else that had transpired since drop-dead gorgeous Lady Jane and her spectacular sidekick, Beverly, had turned up at No. 4 METS. Ensign Westly Slade wondered how developments could possibly get any worse. At this early juncture of his Raiding Forces career the one thing he was sure of was the wait to find out would not be overly long. The Frogs all dutifully linked arms and lifted their boots off the deck even though the plane was not towing a glider.

They felt like idiots.

After the C-47 took off Lady Jane and Beverly stood back up immediately commanding the troops' attention.

Lady Jane said, "It is my understanding you lads have not yet completed your class on the jump commands and what to do in response to them. I shall give a brief recital with Beverly acting as my demonstrator. Listen up—there is a practical test in your immediate future!"

Not even men as hard used—some might go so far as to say abused, as the OSS OG Det 3 Operational Swimmers had been since reporting in to Raiding Forces—could ignore those two. Concentrating on what was being said was another matter. The idea of being taught the jump commands while onboard an airplane en route to making your first parachute jump was so preposterous on its face that it was difficult to process. The fact they were being explained by a titled British noblewoman who was going to be the first jumper out the door of the aircraft and her beauty queen assistant only added to the bizarreness of the overall situation.

Could this really be happening?

Lady Jane said, "Captain Reupart shall be the jumpmaster and his two NCO instructors shall be serving as his assistants. Initially he shall issue three warnings. First is the ten-minute warning which is more in the way of an announcement. The red light over the door comes on at that time. Nothing is expected of you other than to know the aircraft is ten minutes out from the drop zone—a signal for you to dial in and focus on being mentally alert to everything taking place."

Beverly held out both arms with her fingers splayed and in a loud vigorous military manner gave the command, "TEN MINUTES!"

Beauty queens and "vigorous military manner" were such polar opposites it was hard for the Frogs to fathom—this could not actually be taking place. It had to be some sort of test. They would call it off at the last minute. Det 3 had experienced their share of those kinds of drills in the last two years designed to identify and weed out the weak.

Lady Jane said, "Demonstrator recover. The next command will be the six-minute warning which is self-explanatory."

Beverly stamped her boot on the deck of the aircraft and held her arms outstretched displaying six fingers, "SIX MINUTES!"

Lady Jane said, "Now the drill shall pick up momentum. The series of Jump Commands commence immediately. Be ready lads—there will be only short pauses between commands and no time for questions and answers."

Beverly held both arms straight out palms up then raised them toward the roof of the plane, "STAND UP!"

Lady Jane said, "Come to your feet. Do a right face toward the tail of the aircraft taking care not to stumble and fall as you execute the movement. The ride can be bumpy due to turbulence caused by hot atmospheric conditions like we experience in this part of the world. You shall soon discover the plane dances from side to side at inconvenient times. Your parachute, weapons and equipment are all bulky and you shall be crammed in tight like sardines when everyone is standing—take care."

"HOOK UP!"

Lady Jane said, "This one is a little more complicated. In a moment Beverly and I shall work our way down the aisle demonstrating how you are to perform this step. You hook the snap link attached to the end of the yellow static line you shall find running over your right shoulder snapped to the carrying handle of your reserve parachute, to the steel cable running the length of the roof. Remembering to do so *away* from the skin of the aircraft."

This could be confusing. The exit door the Frogs would be jumping from was on the *right* side of the C-47 when you were facing the tail as the jumpers would be. However, it was on the *left* side of the fuselage of the aircraft. Lady Jane decided this was not the time to explain the distinction.

"Then without any additional command from the jumpmaster, insert the safety wire. You shall find it dangling from the snap link on a nylon string. It goes through the tiny hole at the base of the snap link. Be advised this can prove challenging with the aircraft buffeting and swaying causing you jumpers to be thrown up against each other in the close quarters. Once you have the wire pushed through the hole, utilizing your index or trigger finger, bend it down on the far side so the pin will not fall or be pulled out—extremely important.

"Next you form a half loop with the yellow static line and hold it in your right fist. It is imperative you grasp both sides of the static line's loop for dear life. By holding it in this manner you prevent the possibility of accidentally deploying your parachute onboard the aircraft in the event the plane hits an air pocket and you are thrown around or fall on the deck. Understand, your main or reserve deploying onboard is a life-threatening affair that could result in death or injury. It could cause the plane to crash. Watch closely as Beverly demonstrates how to perform this step."

At this point the Frogs were paying rapt attention to the instructions.

Lady Jane said, "Once you are standing up and hooked up, Captain Reupart's two assistant jumpmasters shall begin working their way down the stick checking to make sure everyone has accomplished these two tasks correctly. Normally there would be several additional jump commands but today those have been eliminated for the sake of brevity. Besides, as long as the airplane keeps flying and your parachute opens when it is supposed to, who cares if we leave a few things out?

"I shall not tell if you promise not to."

This drew an uncomfortable chuckle from the Frogs.

"The assistant jumpmaster's inspection shall be brief by necessity because at that point the Time to Target clock is ticking. At the appropriate moment the jumpmaster gives the command . . ."

Beverly stuck out her arms holding up one finger on each hand, "ONE MINUTE!"

Lady Jane said, "The final preparatory command shall be . . ."

Beverly shouted, "STAND IN THE DOOR!"

Lady Jane said, "That one is only meant for me. When I take up my position in the door you close up tight on me—that is important for a fast exit. The next command you hear will be . . .

Beverly shouted, GO!"

Lady Jane said, "As explained, Captain Reupart is the jumpmaster today. I lead the stick, Jumper Number 1, followed by Beverly Jumper Number 2, Ensign Slade is Jumper Number 3 with the rest of the stick following in hot pursuit. I want you lads to exit this aircraft like you are trying to tackle me. Normally I would say if you catch me you can have me—but with the senior Pathfinder on the DZ today being my fiancé, Colonel Randal . . . perhaps you can have Beverly."

This *did* get a laugh.

"Now she shall demonstrate the finer points of the "Airborne Shuffle..."

Ens. Slade was wondering if he was hallucinating or possibly asleep with his eyes wide open having a nightmare. It had happened to him before, during some of the training evolutions during Hell Week with the Navy's Underwater Demolition Teams that involved, among other things, sleep deprivation.

Learning how to parachute out of an aircraft-in-flight while winging your way to the Drop Zone to jump out of said airplane was taking on-the-job training to a new level. Had to be dreaming with his eyes open. No way this was real. He would wake up any minute now.

Lady Jane called out, "Is everybody happy?"

It was not really a question.

The Operational Swimmers of the Office of Strategic Services Operational Group Detachment 3 all lied in the approved Airborne manner, "HELL YES!"

Lady Jane knew a thing or two about the art of military leadership.

CAPTAIN BILLY JACK JAXX REMAINED IN HIS POSITION ON the roof of the building across the street from the 1st Mountain Division Detachments Command Post. The two dead Gebirgsjägers lay in the street. No one had come to check on them. He could remain in the position to observe so long as the Germans did not show up and attempt to maneuver on him. If there were Greek civilians in any of the other buildings in this part of the village they were not showing themselves.

The place was like a ghost town.

Capt. Jaxx performed a mental head count. He had shot the feldwebel and two Jägers back at the cave. He had shot the two at the church on the hill. Next he shot the two sunbathing nudists. He took out the four Germans in the CP. And then shot the two 1st Mountain Division troopers in the street who had come to investigate.

By his original estimate based on experience raiding other islands there would likely be no more than a squad of German troops present on this one. The problem was the TO&E of a squad on paper was one thing. The actual size of one in a division that had been in a large number of hard-fought campaigns—to include serving on the Russian Front in the dead of winter—was another. Chances were a squad of the 1st Mountain Division would have fewer than the ten men found in the standard line infantry Wehrmacht infantry squad.

By Capt. Jaxx's count, he had already taken out more Jägers than he had initially estimated to be on the island. What did that mean? Possibly the 1st Mountain Division was so decimated by all their fighting that the Germans had to combine two understrength squads to have enough troops to station on this island. If that was the case the detachment would consist of an abnormal number of troops and he might have already killed every Nazi in the garrison. Or maybe not.

Capt. Jaxx decided to go find out. He slung his arms and climbed back down off the roof of the building. Time to go hunting.

Capt. Jaxx had hunted men since he was in junior high school—getting his start serving warrants as an unpaid reserve deputy for his grandfather, the county sheriff. To be a successful manhunter you needed to know the tendencies of the men being hunted. In his large West Texas county, where there were more cattle than people, those fell into two categories: local outlaws who knew the territory and Midwestern criminals on the run from their latest bank job hoping to hide out in the boondocks until the heat cooled down. Being sophisticated big-city felons, they believed it would be easy to outsmart the dumb hicks they expected to find in local law enforcement—that was a mistake.

His grandfather, Sheriff Marlboro Jaxx, had served in Company D of the Texas Rangers during the Mexican Incursion. Compared to Pancho Villa's revolutionary bandits he dealt with on the border, tenderfoot gangsters from Chicago or Kansas City were easy prey. The Rangers had never heard of a "writ of habeas corpus" or never read the Fifth or Fourteenth Amendments so they did not know much about the concept of "due process." Capt. Jaxx soaked up his grandfather's crime-fighting techniques. The lesson most applicable to his current situation was criminals generally follow a pattern. They have predispositions that make them predictable.

When stationed on small Aegean islands performing occupation duty, Nazis on any given island tended to spend their days the same ways: 1. Sleep with the best-looking woman/underage girl they could find; 2. Get drunk in a local tavern commandeered for their private use; 3. Avoid work details.

The Germans never showed any interest in trying to win the hearts and minds of the local citizenry. The Nazis had zero interest in civil affairs. Other than how they applied to tendency number one.

What would his grandfather do? Easy. If he were down on the border back in the day, the Sheriff would have found a local confidential informant (CI) and if the informant did not want to talk, well . . . he would.

The situation now was different. His grandfather spoke Tex-Mex. Capt. Jaxx did not speak Greek except for the few words Alex "Cat" Gataki had taught him, but those were of limited military value.

Besides, the locals were laying low and there was still the question of why the two fishermen had guided the Germans to his cave that had not been resolved. Not to forget the abbot who had lied to him about Nazis being present at the church—there was that too. Were the Greeks simply exhausted from being occupied by the Italians, then by the Germans, and now just wanted to be left alone? They would not be if caught aiding an Allied soldier.

Or was the whole island collaborating with the Nazis?

Reflecting on his three-item list of the Nazi occupiers' tendencies, Capt. Jaxx came to the conclusion getting drunk in a bar in the middle of the day might be frowned on by senior management. Officers and NCOs in all armies tend to prefer their troops be sober during duty hours. On the other hand, it also occurred to him that from the EM's perspective, numbers one and three on the list could be accomplished at the same time.

Waldo Treywick loved to quote from the business administration correspondence course Lieutenant General "Geronimo" Joe McKoy took back in the day when he was staking out the Mexican border: "The problem is the solution." And that statement had, in fact, proven to be true in a great many instances when put to the test. Doctor Layton Winthrop improved on it by putting forward Occam's razor: "The simplest solution is best."

When Capt. Jaxx had asked the professor what Occam's razor meant exactly, the answer was, "If you hear hoofbeats, think horses, not unicorns." Which was pretty much nothing more than another take on Raiding Forces Rule 2: "Keep it short and simple."

All of this was a lot to process. Reflecting on his situation the thought occurred to Capt. Jaxx to go grab the first Greek he could find and if the man was not bilingual he could beat the information out of him until he learned English. It might work.

The problem was Colonel John Randal was a vocal proponent of never alienating Greek islanders when at all possible. Which meant Capt. Jaxx was back to square one. He did not know the strength of the enemy forces he was facing. He did not know how to establish the number. And he did not have a clue what to do next.

Being all alone, out of ideas, and since no one would ever know he did it, Capt. Jaxx said out loud, "The problem is I need to know how many Germans are on this island."

Instantly the solution popped out as crystal clear as if The Great Teddy had waved his magic wand—"Hey, Presto." Waldo Treywick would have been proud. A duty roster would be posted in the 1st Mountain Division Detachment Command Post. All senior NCOs kept them. Capt. Jaxx could not speak or read German but he could count.

Reasonably sure he would find what he needed attached to a clipboard hanging on a nail near one of the desks, he headed back to the CP.

Was this "The Problem is the Solution," Occam's razor, Raiding Forces Rule 2 or a combination of all three? Capt. Jaxx did not really care. He could ask the professor later.

Right now all he wanted was a number.

COLONEL JOHN RANDAL WAS SITTING ON ONE OF THE local Castelrozzo mules as it worked its way up the escarpment to the plateau where the Operational Swimmers of the Office of Strategic Services Operational Group Detachment 3 would be making their inaugural parachute jump. Rita Hayworth and Lana Turner, the Zar priestesses/Kit-Kat Club feature dancers were on mules following him. Lieutenant General "Geronimo" Joe McKoy and Waldo Treywick were trailing behind the girls leading a string of additional animals for Major the Lady Jane Seaborn and Beverly Blackwell to ride. And to haul any Frogmen injured too badly on the drop to make it back to ABCHQ under their own power. It was almost like old times in Abyssinia except Happy was trotting alongside and Jack's Pack was marching behind. The boys were along to help recover the parachutes as the jumpers came in. They had spent the morning being taught how to work a Quick Release System in the event any of the Frogs had forgotten the drill and was being dragged.

Lieutenant Chase Starrett had preceded Col. Randal's party up the escarpment to organize the Drop Zone Support Team (DZST). It consisted of a hand-cranked generator and a BC 191E radio to communicate with the lone TTC C-47 troop carrier inbound with OSS OG Det 3 onboard. Today he was the DZSTO—pronounced "Ditz—O," meaning Officer-in-Charge of the Drop Zone Support Team. Col. Randal was giving Lt. Starrett as much responsibility as he could pile on which was what he did to talented junior officers he was grooming for bigger things—Major Butch "Headhunter" Hoolihan and Captain Billy Jack Jaxx being prime examples of past cases.

Onboard the inbound C-47 Dakota serial #42-24335, USAAF 50th Troop Carrier Squadron, Troop Transport Command, Captain Mike "Mad Dog" Reupart stood up in the tail of the aircraft and held out both arms with his fingers splayed. "TEN MINUTES!"

Ninety minutes prior to this heart-stopping announcement, Capt. Reupart and his two assistant jumpmasters had moved to the front of the aircraft and began chuting up the jumpers one at a time, working their way from front to rear of the compartment. This was a time-consuming process and could not be rushed. The Frogs had never put on a parachute before which complicated the process. However, the Operational Swimmers had so much previous training of various types they knew not to fight the problem. That made the exercise easier for the assistant jumpmasters.

Major the Lady Jane Seaborn and Beverly Blackwell had earlier worked their way down the stick of jumpers explaining how to hook up, how to form a loop with the yellow static line so the paratroopers could hold on to it when the snap link on the end was hooked on to the steel cable that ran the length of the aircraft along the top of the fuselage. This was important. When grasping the loop the pressure had to be on a portion of the static line that was hooked on to the cable and *not* to the section that ran over their shoulder into the parachute pack. That prevented any possibility of accidentally activating the main chute onboard the aircraft in the event the jumper was knocked off his feet by the wind gusts that could be expected to buffet the aircraft—sometimes violently—when they were standing up hooked up.

And the women demonstrated how to insert the safety pin into the tiny hole on the snap link and lock it down. The two had taken their time making eye contact with each Frog to establish communication and ensure the instructions were fully understood. The OSS OG Det 3 operators were arguably the most overtrained individuals in the U.S. military. The men understood they were receiving first-class individual instruction from two members of Raiding Forces who were going the extra mile to ensure they knew exactly what to do when the time came. It was clear Lady Jane and Beverly had their best interests at heart. They liked that. The Frogs had been questioning the "best interest" part.

Raiding Forces clearly took the concept of not being "for the weak or faint of heart" to the max.

Back in the tail of the Dakota immediately following the ten-minute warning, Lady Jane stood up to address the Frogs. "When you come in to make your PLF I want you lads to make two fists covering your face to include your eyes. But keep them open at all times. You shall need your peripheral vision. Place your forearms tight together touching each other from your fists all the way down to your elbows and keep your head down, chin on your chest. Is that clear?"

"CLEAR!"

"Keep your knees bent from the moment you make your exit. Just before you touch down rock your boots slightly so you know for absolutely certain they are NOT locked—is that clear?"

"CLEAR!"

"One last thing lads—if your main parachute does not open after you reach the four thousand count, staying in the tuck position, immediately reach down with your right hand, grasp the metal ripcord handle on the reserve strapped to your chest and jerk it as hard as you can. It will come off in your hand. Never fear . . . that is what it is designed to do. You have not broken it—is that clear?"

"CLEAR!"

"Next, and this is essential, drop the handle, reach in with both hands and *pull* out the reserve parachute canopy. Then toss it away from your body downwind. It will not automatically deploy. That part is up to you. If the canopy blows back in your face keep throwing it out until it catches the wind and pops open. You have to keep working the problem—is that clear?"

Det 3 was already bordering on mass cardiac arrest. Her instructions did nothing to reduce the stress level. "Work the problem" while screaming to your death at some unknown terminal velocity—estimated to be around 150 miles per hour? Oh yeah, that sounded like a good idea. How were they supposed to know which direction was "down wind"? The thought of jumping out of the airplane was a lot more terror-inducing than the run-up to their first combat mission had been. It was known Lady Jane had a sense of humor. Was it possible she could be joking about deploying the canopy by hand?

"CLEAR!"

Unknown to the Frogs, Lady Jane intentionally left out a few details—like the fact the reserve's canopy could fall out and wrap around their ankles if they failed to deploy it by hand fast enough. Or how the canopy could blow back and become entangled in the lines of the main chute over their heads—out of reach. Or that when a reserve chute did open it frequently corrected whatever malfunction the main chute was experiencing and now the jumper was dangling under two canopies.

While two parachutes might sound like a good thing, it was not. The reserve chute has shorter lines than the main canopy. That meant it was likely to steal the air from the main chute causing it to collapse. In that event, two things could happen and they were both disadvantageous for the jumper. The main might collapse, fall down, land on top of the smaller reserve canopy and cause it to deflate. Or the main could fall past the reserve canopy, reopen, and steal its air, causing it to collapse. Then the reserve might fall down, reopen and steal the air from the main again etc. It was not desirable for a jumper to be thirty or forty feet in the air while experiencing the seesaw effect of having the canopies stealing the air from each other inflating and deflating with neither chute, main or reserve, fully open at the moment of coming in to perform their PLF.

This, she decided, was information best left unsaid.

Lady Jane said, "Most malfunctions are caused by "bad" exits which usually translates into *hesitant* exits—you lads chase me out the door like The Hound of the Baskervilles."

Capt. Reupart shouted, "SIX MINUTES!"

Now time compressed. Events taking place onboard the aircraft suddenly switched from agonizingly slow to fast forward. For novice jumpers now the experience was not unlike being a passenger on a runaway train—trapped in a time warp, watching events unfold with no control over anything but knowing it's not going to end well.

"STAND UP AND HOOK UP!"

On the DZ, Lt. Starrett pressed the "push to talk" switch on the BC 191E radio's handset to respond to Captain Richard Holden's query about wind conditions. “Everything is copacetic.” Which is paratrooper speak for “Good to go. Bring it.”

Captain Billy Jack Jaxx had once advised Lt. Starrett there were three things expected of a Raiding Forces officer: 1. Lead from the front; 2. Look cool while doing it; 3. Always be cool.

The trick to laid-back radio transmissions is to keep them as brief as possible and sound bored—even if the Second Battle of the Little Bighorn is raging around you.

Lt. Starrett ordered, “Pop smoke.”

Private First Class Norvel “Horn Dog” Hansen was standing by with a smoke grenade ready. He pulled the pin and tossed it on the ground.

Lt. Starrett radioed, “Smoke out—authenticate.”

The smoke swirling from the grenade was a distinctive violet shade—a tint favored by the British Airborne. The U.S. troops in Raiding Forces described the color as purple. It paid to be careful when dropping paratroops or making resupply drops during daylight hours. The other side was known to put out its own smoke in an attempt to lure the pilots into mistakenly green-lighting the jump on the wrong DZ.

That was not going to happen on Castelrozzo today since there were no longer any Germans present on the island. But it was not a bad idea to follow procedure every single time to develop mental muscle memory. It is commonly said you fight how you train, which is not always true, but it’s one of those catchy concepts that it never hurts to put into practice on training exercises like an administrative parachute drop.

Capt. Holden drawled, “Goofy Grape.”

Not to be one-upped in laid-backness, Lt. Starrett pressed the Push-to-Talk switch. “Go for it, birdman.”

Capt. Holden said, “Batter up.”

IN THE SHAMBLES OF THE SEMI-DESTROYED COMMAND Post, Captain Billy Jack Jaxx found what he was looking for on a clipboard hanging from a nail behind what he guessed was the duty desk—right where he expected it to be. There was a roster starting with the feldwebel . No officer was listed or at least he saw no prefix of a rank he might have recognized as an officer. There were sixteen names.

One thing Capt. Jaxx knew. Germans were efficient. If that 1st Mountain Division Detachment roster said sixteen men were present for duty on this station there were sixteen men. He had only accounted for fourteen of them.

What to do?

Even tiny islands had police departments. When the Nazis occupied a place the first course of action they took was to disarm everyone to include the local police. They would leave the police officers on the job to maintain order—without firearms. The Germans did not want any unsecured weapons in the hands of the populace law enforcement or not if they knew the Greeks hated them. Although the hating/resisting part on this island was not entirely established in Capt. Jaxx's mind.

However, while the locals' loyalty had not been established to his satisfaction it was a safe bet there were not a lot of unsecured guns floating around. So if he walked into a police station, odds were no one was going to shoot at him. And the policemen might be able to tell him where to find the unaccounted for Germans.

The question was how to find the police station. During his hide-and-watch surveillance, Capt. Jaxx had not spotted a single policeman patrolling the village. He did not have a clue what the Greek word for "police" might be but he had encountered police stations on other islands and might be able to recognize the word. A long shot but he did not have any better idea.

The good news was he had dramatically reduced the odds against him. The bad news was he was still outnumbered two to one by highly trained, battle-hardened, extremely dangerous enemy soldiers who were out there somewhere armed to the teeth. Capt. Jaxx started walking down the street toward the turquoise bay he could see in the distance. There was a dock. No one was in sight; however, the caiques and schooner were still tied up. If he knew where he was he could hijack one and sail away. Only he had no idea where he might be more specific than planet Earth—which is not a good starting point for aspiring navigators.

Escape by sea was not an option.

It was beginning to dawn on him that he'd had a long day. It was imperative he think things through and not simply react. He might be a little tired. There was a trail of dead Germans strung out behind him. While that was good, Capt. Jaxx still needed to take down the remaining two Jägers. And not make any mistakes when doing so. That might be easier said than done.

He came to a snow-white, one-story building with *αστυνομία* painted over the door. There were bars on the windows. It had the appearance of a police station you might see in a small Mexican border town. The place looked like the other police departments he had come across in the islands.

Capt. Jaxx edged his way down the side of the wall and made a quick head check through a window. Inside a middle-aged man was asleep on a sofa so old it should have been in a museum. He was snoring. The sleeper was wearing a stained khaki short-sleeved shirt that might or might not have been part of a policeman's uniform. There was no way to tell.

Capt. Jaxx unslung the SVT-38 rifle. Being as quiet as possible he removed the magazine and ejected the round from the chamber. Then did the same with the two Steyr MP-34 submachine guns. He leaned all three against the wall. He did not want anyone coming up behind him to have access to a loaded weapon.

Then Capt. Jaxx took the P-38 Walther pistol out from the back of his jeans where he had been carrying it. He cracked the slide enough to visually see if there was a round in the chamber—there was. Just like the last time he checked. The P-38 was the most modern handgun design in the world due to its single-action/double-action trigger mechanism—meaning the weapon did not have to be manually cocked for the first shot but you could if you wanted to. Doing so made it single-action and that made the trigger pull lighter. For double-action, provided there was a round in the chamber, all you had to do was to pull the trigger.

Capt. Jaxx cocked the exposed hammer with his thumb and held the weapon down beside his leg. He did not like the Walther P-38. Its double-action trigger pull, which was the design feature the handgun was famous for, was between sixteen and twenty pounds, depending on the individual weapon. And it did not have a clean break. Single-action, meaning starting out with the hammer cocked, made the trigger pull lighter but it was still over six pounds and gritty.

On the .45 Colt Single-Action Army revolver he grew up shooting, Capt. Jaxx preferred a trigger pull adjusted to two pounds—he fired it with the tiniest tip of his finger. His Colt Government Model 1911 38 Super fighting pistol's trigger, tuned by Lieutenant General "Geronimo" Joe McKoy, was three pounds and broke like a glass rod. Many knowledgeable gun handlers would not be comfortable with those trigger pulls, believing they were too light for safety.

Capt. Jaxx did not carry a pistol to be comfortable.

Last thing before making his entry was to remove the camouflage 1st Mountain Division smock. He was walking through that door as a U.S. Army paratrooper—jump wings on his chest and a Raiding Forces flash on his shoulder. Inside, the sleeping police officer jerked awake when Capt. Jaxx kicked the couch. He was startled, possibly afraid, and needed a few moments to regain his composure. The officer finally realized he was being confronted by an Allied soldier though he had never known of one wearing blue jeans before.

In reasonably good English the policeman said, "How are you today?"

"I'm fine, thank you."

"Have you come to liberate Hydra?"

"Actually I'm looking for two Nazis."

"You will find them in our holding cell. Down the hall through the doorway. The German sergeant in charge of the Wehrmacht detachment despoiling our island ordered the two confined earlier this morning for turning up in formation for roll call drunk and disorderly."

Capt. Jaxx eased down the hall and looked through the open doorway. There was a single jail cell in the back room. Two Gebirgsjägers were sitting on the edge of their bunks with their heads down smoking cigarettes. They definitely appeared hung over. A condition commonly described back home in Texas as "Rode hard and put away wet."

He shot each of them three times.

THE FROGMEN OF OSS OG DET 3 WERE STANDING FACING the rear of the aircraft, hooked up, having been issued the one-minute warning. It did not look like the drop was going to be called off. "Prejump jitters" were in full force and effect due in small part to the wise-cracking rigger Captain Karen Montgomery. When the Frogs drew their parachutes at the Departure Airfield she responded to the inevitable question rookie paratroopers always ask prior to their first jump, "How long have you been packing parachutes?"

Capt. Montgomery's standard issue response was, "My first day of on-the-job training. If it does not work as described bring it back and I shall give you another. Have a nice jump."

Not what anyone wanted to hear.

The two assistant jumpmasters quickly made their way down the stick conducting one cursory final safety check—not normal procedure—mostly done to bolster the Frogs confidence. If they found any deficiency it was too late to correct it. Major the Lady Jane Seaborn was standing in the door, her right boot slightly forward with the tip of the toe partially outside, both arms outstretched angled slightly down outside, palms flat against the skin of the C-47 Dakota to give her leverage for her exit, knees flexed. She was ready to launch on the Command "GO."

Lady Jane liked this part, looking out over the horizon a thousand feet up, C-47 thundering toward the drop zone, wind tearing at her battledress, riding with the swaying and buffeting of the aircraft. Standing in the door waiting to go is when you discover airplanes do not fly in nice, neat straight lines at all times.

Captain Mike "Mad Dog" Reupart, her instructor ages ago when she and Captain Pamala Plum-Martin went through their parachute training, had his arm braced across the door to hold her in until the green light. Lady Jane was an experienced parachutist. She and her Royal Marines, FANYS and Wrens made weekly training jumps on this very DZ out of the Hudson or one of the Walruses. Nevertheless, as was frequently said in Raiding Forces, "Why take a chance?" Mad Dog did not want her to exit prior to the green light due to some miscommunication.

Capt. Reupart noted not for the first time that Lady Jane was even more beautiful when she was tightly focused like right now, leaning over his arm and looking up ahead for the smoke put out by the DZST. Making her own estimate of the situation. Coming up fast the coastline of Castelrozzo was the Initial Point (IP). It flashed past below. The Release Point was an imaginary line five seconds later when the green light would come on based on wind conditions on the drop zone—different for every jump.

A stream of tracers from a pair of 7.92 mm M17 machine guns mounted in the wings of a Luftwaffe Ju-87 Stuka dive bomber flashed past out of nowhere and ripped into the C-47's engine on the left wing directly in front of Lady Jane. Two German dive bombers on an anti-shipping patrol were returning to base on Rhodes without success when they spotted the unarmed Dakota headed toward Castelrozzo and thought it would be easy prey.

That was a mistake.

After the war moved on from the desert to North Africa, Sicily and then Italy, there was a surplus of antiaircraft batteries left behind in Middle East Command. Vice Admiral Sir Randolph "Razor" Ransom had commandeered all he could for Small Raids Inc. He ringed Castelrozzo with AA, cramming in as many air defense guns per square foot on the perimeter of ABC as there were on Malta, an island fortress frequently described by the world press as the most heavily defended place against air attack in history.

The AA gunners immediately opened. One of the British Quick Fire 3.7 in.—an equivalent gun to the US 90mm—hit one of the two Luftwaffe aircraft with its initial barrage. The dive bomber blew up. Seconds later the second German aircraft was shredded by a crossfire of Vickers 50 cal. heavy machine gun (HMG) rounds from a pair of Royal Navy HMGs mounted in four-gun battery antiaircraft mounts. It winged over, crashed, and exploded, almost taking out the DZST.

Simultaneously, the engine on the C-47 Dakota that was hit began trailing flame.

Lady Jane did not panic, hesitate or wait for Capt. Reupart to assess developments. She leaned back inside the cabin and shouted, "FOLLOW ME LADS—LET'S GO!"

Then she was out the door with Beverly Blackwell so close behind as to be physically touching the parachute pack on her back with the

Frogs charging after her exactly as instructed . . . like a pack of "The Hound of the Baskervilles." None of them were thinking about having their way with the Texas beauty queen if they caught her. Capt. Reupart and his two assistant jumpmasters followed them out. The USAAF crew chief and the radio operator came right behind.

Down on the DZ, looking up watching the stick spill out of the Dakota, Waldo Treywick said, "Frogs may've just set a speed record for unassing a C-47."

The burning engine was causing the Dakota to vibrate violently. Before the TTC pilots could feather the prop and shut it down, it fell off completely. The engine slammed into the DZ catching the grass on fire. The sudden loss of the weight on the left side of the aircraft, combined with the full power of the right engine, caused the C-47 to start to roll over. The pilots cut the power on the right engine to bring the aircraft back to wing level. It required both of them struggling with the right aileron and rudder to keep from losing control.

At this point in what was supposed to be a simple training mission—drop a few paratroopers and then back to the officer's club for drinks—the plane was going down and the pilots had no option except to ride it in. In addition to making a full-on crash landing there was also the not-so-small matter of starting to run out of island to contend with.

Down below in the vicinity of the DZST a full-scale rodeo was in progress. The mules did not take kindly to the sudden cannonade of heavy caliber AA weapons fire. And they definitely took exception to the Ju-87 crashing fifty yards away and its 1,100-pound bomb exploding. USAAF doctrine called for airstrikes not to be brought in closer than one mile of friendly forces. Whoever wrote that directive must have been some armchair commando who had never actually seen an airstrike or needed one called in. Air/ground support is generally brought in close no matter what the regulation says because that's where the bad guys are.

A lot of things can go wrong when airstrikes are called in close to troops. Bombs contain high explosives and create shrapnel. They are dropped by a pilot flying at low level who may be taking ground fire at the time, which can be distracting. He might be flying his first combat mission. Or he could have a hangover from a hard night of partying at the O-Club. It is an inescapable fact *some* pilot has to graduate at the bottom of his class.

In this case, it was a fully armed Stuka dive bomber crashing out of control and it did not matter what the pilots class standing had been. No one was harmed when it pancaked—other than the German and his navigator/radio operator/rear gunner. Nevertheless, 1,100-pound bombs are big, noisy and dangerous. The detonation was end-of-the-world type deafening. It seemed to rattle the whole island. And the razor-sharp pieces of shrapnel screaming past did nothing for morale—man or beast, especially the mules.

Colonel John Randal had been sitting easy on his animal with one leg—the one he had injured—cocked over the pommel of the English-style saddle, resting it. When the antiaircraft fire started the mule went berserk. One second Col. Randal's mild-mannered steed was standing peaceably and the next it was sideways in the air all four legs parallel to the ground. Not unlike the attitude a trainee about to learn the joys of a PLF on a Swing Landing Trainer might find himself in just before the instructor shouts "Land!" and lets go of the rope. When the mule came down he landed stiff-legged, made a few head-down crow hops, then kicked for the moon with both hind legs.

Col. Randal never had a chance. He went sailing over the animal's head. There were a lot of bad landings in progress on the DZ but none worse than his. To his credit he did not actually lose consciousness when he slammed into the ground .And he managed to never let go of the reins.

Happy raced over to lick his face.

Lady Jane floated down as light as a feather, made a stand-up landing against regulations prohibiting them, and dropped her X-type parachute. She ran up, shooed Happy out of the way, knelt down beside him and sounding like a mother consoling a small child said, "Did you fall off your mount?"

Col. Randal knew he was never going to live this day down.

Lieutenant General "Geronimo" Joe McKoy trotted by, "These jacks are some sure 'nuff skittish saddle stock, John. Ain't suitable for Sunny Brook Farm-type trail rides. That's for dead certain."

Which did nothing to make him feel any better.

Waldo rode up, leaned down, and handed him a cigar, which he stuck in his front teeth, glad to have it.

Beverly rushed over with Rita and Lana, clearly concerned. "That's a very tall mule to fall off of, Johnny. Are you sure you're OK?"

Wanting to change the subject, Col. Randal gave a totally unnecessary order since it was already being carried out. "Get everyone out there to help Jack's Pack knock down those parachutes."

Ensign Westly Slade was coming in for a landing. He had been briefed that front PLFs were as rare as the dodo bird, yet here he was making one. At this point Ens. Slade went blank and totally forgot everything he had been taught about parachute landing falls which, in his defense, was not very much. He did not have his fists over his face, his forearms were not squeezed together, his elbows were not tucked in, his chin was not down on his chest, he did not rock his knees prior to landing and he did not make any lightning-fast split-second decision to conduct either a right or left forward PLF.

He just burned straight in.

There may have actually been some *Blood on the Risers* on this drop zone. When Ens. Slade hit the ground he had his feet and knees together. That turned out to be a mixed blessing. While he did not break his ankles or shatter the bones in his legs as warned, the Ensign landed stiff as a board. Instead of rocking his boots as he came in, Ens. Slade had instinctively locked his knees in anticipation of what he rightly suspected would be an experience that resulted in sudden pain. The forward momentum of hitting the ground on the balls of his feet at speed with his knees locked caused him to flip. The end result was a spectacular three-quarter full front gainer to land flat on his back.

From where Col. Randal was, it appeared Ens. Slade might have bounced. Had to hurt. One of Jack's Pack ran over, grabbed the sailor's harness, and pulled the dazed Frogman to safety moments before the grass fire caught up to him. Ensign Slade's PLF, or lack thereof, improved Col. Randal's morale more than a little and he did not feel one bit guilty about it—not even slightly.

Overhead an Operational Swimmer was coming in upside down. He had apparently made the mother of all bad exits, managing somehow to dive head-first through his parachute's lines as they deployed and got one of his boots entangled. This created a malfunction known as an "inverted canopy" which is exactly what it sounds like. His parachute was turned inside out. Not a problem, it still worked. Fortunately because of all their hard Stateside training, OSS OG Det 3 personnel were Olympic-class athletes. Seconds before piledriving headfirst into the ground, with likely catastrophic consequences, he bent at the waist performing a sort of aerial high-speed sit-up and actually managed to land on his fifth point of contact—the meaty portion of the back. The Frog timed it perfectly. In fact, he tagged four of the Five Points of Contact only failing to land on the balls of his feet. All in all not a bad PLF—just in the reverse sequence.

Jumpers were coming down. The wind was whipping the flames. Parachutes were catching fire.

Then the C-47 Dakota crashed.

6

A TOUGH OUTFIT WITHOUT THE SWAGGER

COLONEL JOHN RANDAL PULLED HIS REINS OVER HIS mule's head preparatory to mounting. The sudden antiaircraft fire, one of the Ju-87 Stukas blowing up approximately a thousand feet overhead, the other crashing, exploding, nearly taking out the DZST, the C-47's burning left engine falling off causing a grass fire and strange creatures drifting down out of the sky under giant silk mushrooms all happening pretty much at the same time contributed to the animal shying away violently. He did not want any part of what was taking place.

Col. Randal had banged up his injured leg when the mule pitched him off, which slowed him down now. Getting aboard the reluctant animal was complicated by the lack of a saddle horn he could grab onto. Instead of a Western pleasure-style saddle, the mule had been fitted with the Greek version of an English riding saddle with no horn and tiny little metal stirrups, which did not make getting aboard any easier either. When he finally managed to get a grip under the pommel, the animal quit shying sideways and began circling hard to the left because Col. Randal had the reins pulled in tight jerking his head around. But the mule started to crow hop trying to get away. Two quick running steps, one of which was excruciatingly painful, and Col. Randal swung up on his back and they were off and running—forget the stirrups. Once given his head, the long-legged animal was really picking them up and laying them down.

Col. Randal was racing to rescue the two C-47 pilots—the mule, he was getting the hell out of Dodge.

Lieutenant General "Geronimo" Joe McKoy and Waldo Treywick came galloping right behind as did Major the Lady Jane Seaborn and Beverly Blackwell on the saddle animals brought up the escarpment for the two of them to ride back to ABCHQ. Rita Hayworth and Lana Turner were pacing them riding like the wind. Lieutenant Chase Starrett and Private First Class Norvel "Horn Dog" Hansen stopped what they were doing, mounted up, and took off at a dead run to catch up. The C-47 made a wheels-up belly landing, skidding for two hundred yards before coming to a stop. While it had not caught fire or exploded, either or both of those things could happen at any second. The pilots were still inside.

When Col. Randal arrived at the crash site he swung out of the saddle but his mule shied away causing him to come down hard on his injured leg. It buckled. He ended up on the ground again as Lt. Gen. McKoy and Waldo bailed off their mounts and ran past, charging through the open exit located in the tail of the aircraft. Lt. Starrett and PFC Hansen rode up, jumped off and dashed inside the Dakota right behind them. It takes courage to run into an airplane that might explode at any second.

Lady Jane and Beverly dismounted and knelt down beside Col. Randal, "I'm all right."

"That would be why you are lying on the grass instead of inside that airplane playing at being a hero," Lady Jane said. "I worried about the OSS lads today. Not you, John."

Col. Randal was right he was never going to live this down.

Beverly said, "Dr. Milam's strict instructions were to stay off your leg for a week. You haven't paid one bit of attention to him, Johnny. You're not setting a very good example for your troops."

That hurt. Col. Randal thought he was setting the example—pressing on through his injury. "Why do you think I was riding a mule? Following doctor's orders."

Lady Jane laughed, "Looks as though your mule has been riding you cowboy."

Having ended up on the ground twice in less than ten minutes in front of witnesses—embarrassing for a former officer in the U.S. 26th Cavalry Regiment—Col. Randal could not think of a single response the women would have paid attention to.

Captain Richard Horton and his co-pilot Lieutenant Malcolm North staggered out the exit door, clearly shaken up by the crash landing. Both men were displaying cuts and abrasions. Neither appeared seriously injured.

Beverly said, "Nice work, boys. Any landing you can walk away from is a good landing."

Capt. Horton said, "Make sure to tell Bronc that."

"I'll do that right after I've finished writing you up for the DFC, Captain—didn't have to turn out like this."

When Lady Jane stepped away to see for herself if either pilot required medical attention, Col. Randal said, "Beverly, when we get back to ABCHQ buy that mule for me."

"Why would you want me to do that?"

"So I can shoot him."

Lady Jane said, "Rita, you and Lana allow these two pilots to hop up behind you. Take them down to see Dr. Milam—be gentle."

Lt. Gen. McKoy came out of the airplane, "We're burning daylight, folks. We've got places to go. Things to do. People to meet."

Col. Randal said, "Roger that. Let's move. Lieutenant Starrett, have the Det 3 personnel assemble. I need to speak to the men first."

"Yes, sir."

Lt. Starrett and PFC Hansen mounted up and loped back to the DZST. Lady Jane held Col. Randal's mule's bridle to quiet the still agitated mule while he mounted. Following at a sedate walk, he said, "Beverly, you're flying soon. Anything you need to do prior to?"

"Nothing more than my preflight."

"Get some rest then. You've got a long mission ahead of you. I need you at your best."

"I managed a nap on the plane."

"General, how about you?"

"Just to check out our casualty rate after Dr. Milam gives the Frogs the once-over. Worst jump I ever saw. We'll have to figure out how many a' those Frogs we're goin' to need to replace before LD time."

Waldo said, "Could be all of 'em."

To her credit, Lady Jane refrained from saying *I told you so*.

When they arrived back at the DZST Lt. Starrett had the Frogs assembled. The medics were making an initial assessment of jump-related injuries and providing immediate aid for those who required it. Almost every man needed some form of treatment. The good news was after a cursory inspection no obvious broken bones had been found—meaning no compound fractures. Though with as many people banged up, that would have to be confirmed by Dr. Stephen Milam.

Morale was sky-high.

Lady Jane was shocked to hear one of the Frogs say, "It sure is a good thing we had those British X-type parachutes with their gentle opening shock."

She pulled Lt. Gen. McKoy aside. "Why are these men laughing and celebrating after all they have been through today?"

Lt. Gen. McKoy said, "Well, for one, Lady Jane, they didn't get themselves killed. They're on their way to bein' full-fledged paratroopers—they can do this. John's got a magic touch when it comes to handlin' troops. The Frogs don't know it yet but he's about got 'em wrapped around his finger."

Lady Jane said, "I fail to see how that could possibly be, General."

"Those OSS boys rolled in here with big egos and the idea a' showin' us how it's done," Lt. Gen. McKoy said. "Way too much attitude for their own good. John's turned the tables on 'em."

"How is that possible?"

Lt. Gen. McKoy said, "Soon as they got here John sent them all on a live fire combat mission against a hard target, that very night—who does that? No officer I ever served with. Then he sends 'em on another one. Next he packs 'em off to parachute school. Doesn't ask for volunteers—just go do it and right now. Move out smartly, boys. Then without so much as one full day a' Airborne trainin' the Colonel orders the whole shootin' match to load up and make an actual jump out of a aircraft in flight. You can bet that got the Frogs' attention."

Lady Jane said, "Got mine."

"Today's drop 'll be the stuff a' legends. Doesn't hurt the story none that the transport got shot down. The Frogs'll be talkin' about this jump for the rest a' their lives. What they did today'll earn the respect a' our more experienced jumpers and go a long way toward cementing unit cohesion. John's just about got the OSS boys infused into Raidin' Forces culture on his terms and they never saw it comin'— like I said he's got the magic touch.

Lady Jane said, "You believe that was John's plan all along? A unit-building exercise? If that is so, he nearly killed the OSS lads in the process."

Lt. Gen. McKoy said, "Every now and then, Lady Jane, it's necessary to swat a' fly with a' sledgehammer. The idea ain't to kill the fly—it's to get the attention a' all the other flies. Oh yeah, John's had somethin' like this in mind from day one. The Frogs are one a' us now…almost."

"I shall take your word for it General if you are truly convinced these last few days have been a leadership stratagem?"

"Yeah, I am. A beautiful thing militarily speakin'," Lt. Gen. McKoy said. "John ain't as good at it though when it comes to dealin' with you Raidin' Forces women."

Lady Jane laughed. "I thought no one noticed."

"You thought he was bein' too rough on the Frogs?"

"One of John's most appealing qualities is his hard edge. Do not tell him I said so. Makes me feel safe."

Lt. Gen. McKoy said, "Well, you might want to keep in mind your boyfriend's probably got hisself some ulterior motive for the OSS boys he ain't sharin' with anybody right now so don't be shocked when it goes down."

"When you find out what that is, let me know so I can prepare myself," Lady Jane laughed.

Sitting on his mule at the DZST, Col. Randal ordered, "You men fall in around me. Drop your parachutes. They'll be recovered for you."

If the Frogs were expecting congratulations or even an acknowledgment of having successfully completed their first military static line parachute jump or some comment about the C-47 being shot down, they were mistaken. Col. Randal acted as if the drop—to include the shootdown—was nothing more than business as usual. It was beginning to sink in on OSS OG Det 3 this was probably the way things worked in Raiding Forces—a tough outfit without the swagger.

Things came at you hard and fast.

Col. Randal said, “We’re moving out to ABCHQ. A hot meal is waiting in the Other Ranks mess. Then you’ll report to the Aid Station for a more comprehensive medical exam.

“After that I’ll be issuing a Warning Order. As of now Det 3 is on Stand-By Ready Alert. Are you men prepared to accept a new mission?”

This time the OSS Operational Swimmers were only semi-lying. “HELL YES!”

What could possibly be worse than their day so far?

VICE ADMIRAL SIR RANDOLPH “RAZOR” RANSOM WAS waiting for Lieutenant General “Geronimo” Joe McKoy and Colonel John Randal when they arrived back at ABCHQ. The three repaired to his office and shut the door. A late-breaking development that was going to have an impact on the night’s operation needed to be discussed in private.

VAdm. Ransom said, “The submarine *Perseus* Captain Jaxx is aboard has missed its last three scheduled radio checks. The sub has now officially been declared missing. You should be aware in the Navy a missing submarine is presumed lost—but is not classified as such until the fact can be proven by the recovery of some artifact. In many cases that never occurs.”

This unexpected announcement came out of nowhere. The news hit Col. Randal like a body blow. It was the single worst piece of news Col. Randal had received since he started fighting Huk guerrillas before the war started—maybe ever. At least when Lady Jane was shot outside the Gezira Club she had a chance of survival. Jack was dead. Out of his peripheral vision he could see Lt. Gen. McKoy's features were etched in stone.

VAdm. Ransom said, "I withheld the missed checks because there was nothing anyone could do about it and there was always the possibility *Perseus* would eventually make radio contact. My estimation of the situation was and still is the rescue of the NAPRW pilot has such far reaching political implications as to require full attention from Small Raids Inc./Raiding Forces, free from emotional distraction. That shall not occur if word gets out that our most popular junior officer has been killed in action—time enough to grieve later."

Lt. Gen. McKoy said, "What's changed, Admiral, to make you tell us now?"

VAdm. Ransom said, "Possibly nothing—however there has been a development. Understand you need to take what I am about to inform you of next with a healthy dose of skepticism. Our Y-Service wireless people picked up a garbled transmission that claims to have been broadcast by Captain Jaxx."

Col. Randal said, "From the *Perseus*?"

VAdm. Ransom said, "Y-Service claims the message originated from enemy-occupied Hydra Island. While not impossible it is highly unlikely Jack sent it. The signal is suspicious on its face."

Lt. Gen. McKoy said, "What made anyone think it's from Jack?"

VAdm. Ransom said, "The sender gave the Captain's SOG identifier. Then the transmission broke up and contact was lost. Attempts to reestablish it have been unsuccessful. The Y-Service boffins were able to triangulate the general location of where the message emanated but that is all we know at this point. This could be a signals intelligence game the Nazis are running in hopes of tricking us into sending in a rescue party."

Col. Randal said, "What's your opinion, sir."

"Personally, I fear a trap."

Lt. Gen. McKoy said, "What do you recommend, Admiral?"

VAdm. Ransom said, "We need to keep our priorities straight. Take this one step at a time. First let's concentrate on retrieving the pilot on Sikinos. That shall give Signals Intelligence more time to evaluate the radio transmission. Then we can decide on the best course of action for going after Capt. Jaxx on Hydra.

"The main thing is I do not want you to get your hopes up. There is virtually no chance the Captain sent that message. The most reasonable explanation is his body washed up on the beach, the Germans recovered his identity disc, and the Abwehr has a file on him that might contain his SOG identifier. With that information the Nazis would be able to send out an intentionally garbled transmission to lure us into an ambush. OKM, the Kriegsmarine Naval Intelligence Department, runs that sort of SIGNIT game. Same as our side."

Col. Randal said, "OK Admiral sounds like a plan."

Lt. Gen. McKoy glanced sideways at him.

Ignoring the look, Col. Randal said, "General McKoy, you continue organizing the Sikinos operation. Brief me in my quarters as soon as you've finalized the details for the Warning Order. I'll need you to bring me up to speed so I can issue it."

Lt. Gen McKoy said, "Roger."

Col. Randal said, "Admiral, can I count on you to keep me informed of developments—no holding back, sir?"

"Absolutely," VAdm. Ransom said. "We should limit distribution of the intel on the *Perseus* to the three of us for the time being."

Col. Randal said, "Agreed."

He limped out of the TOC and up the stairs to his suite. The leg was beginning to stiffen up. Flanigan was on the duty desk.

"Find Corporal Murphy and have him report to me ASAP if not sooner. Swear Murph to secrecy about coming up here. And tell him to be discreet. I don't want anyone to know he was here."

"Sir!"

"Either of you mention he's been to see me, Flanigan, and I do mean *anyone*, I'll have you both transferred to the Russian Front."

The policeman said, "I was not aware the British or the Americans had troops stationed on the Russian Front, Colonel."

Col. Randal said, "In that case, the Reds could probably use a couple."

"Sir!"

Inside the suite, Col. Randal went to the wall map he had in the small private briefing area and searched until he located Hydra Island. It was tiny. Only one village, Hydra Port, was labeled. Not much else was to be learned about the place from looking at the map. He needed more information so he phoned Dr. Layton Winthrop and requested he come upstairs without saying why other than to use the word "discreetly."

There was a knock on the door. Col. Randal opened it. Lt. Gen. McKoy was standing there. "Sounds like a plan? When you got bucked off your mule did you land on your head, John?

"You're gonna do 'first things first' in lockstep with the Admiral to make a couple a' politicians happy? Leave Jack twistin' in the breeze?"

Col. Randal said, "Hell no, step inside—I'll show you what we're going to do."

FLANIGAN RETURNED TO COLONEL JOHN RANDAL'S SUITE with Beverly Blackwell's navigator, Corporal Tom Murphy, in tow. She called him "Murph the Surf." The New Zealander had been in the LRDG from inception until transferring to Raiding Forces when the unit moved to Oasis X and had a need for experienced desert navigators to guide its gun jeep patrols. When Beverly started flying she selected him to do her navigation.

Col. Randal said, "Flanigan, did you mention the part about Russia to the Corporal?"

Cpl. Murphy said, "He did, sir."

"That order remains in full force and effect until I tell you differently," Col. Randal said. "It includes informing Beverly or Lady Jane. Is that clear?"

"Perfectly, sir."

"And don't you forget it, son that threat ain't no drill," Lieutenant General "Geronimo" Joe McKoy said. "You goin' to have trouble findin' the High Speed Launch tonight, Murphy?"

Cpl. Murphy said, "It's a tricky piece of navigation, General, but if the sailors are where they are supposed to be we can locate the HSL, provided it can turn on lights to guide us in for the last mile if we are under cover of darkness at that point in the flight."

Col. Randal glanced at Lt. Gen. McKoy. Confidence is a good thing. Pinpoint a sixty-three-foot boat in the middle of the Aegean Sea at night—"no problem" might be a little too much self-assurance. But then Murph the Surf had been navigating mobile desert patrols in the Great Sand Sea and Beverly to distant targets for a long time now.

Col. Randal said, "What we're about to discuss does not leave this room, Corporal—clear?"

"Sir!"

"I'll be aboard the Hudson but that is classified," Col. Randal said. "Once we land, the General and his party will transfer to the HSL and set sail for Sikinos to rescue a photoreconnaissance pilot known to be at large on the island. At that point, prior to taking off I'll issue Beverly a change of mission. I want you to act like you're hearing it for the first time—is that clear?"

"It is, Colonel."

Col. Randal said, "Let's take a look at the map. This is Hydra Island. The new orders will be to plot a course from the HSL to Hydra Port—located here. The Hudson will not be landing. At least initially. It will simply overfly the village. Make your approach from the sea. Overfly the village. Do not come in from the land side of the island. Is that clear?"

"Sir!"

"Be ready with your flight plan. I don't want to spring this on you, Corporal."

"I understand, Colonel."

The three gathered around the wall map and looked at the location Col. Randal indicated. Maps are read right and up. The top of a map is always north. Like the majority of islands in the ATO, Hydra appeared as a tiny speck of sand in a big sea. To complicate navigation it was not far from Greece. According to the scale, the island was thirty-seven nautical miles distance from Athens. Overshooting in the line of flight from the rendezvous with the HSL would put the Hudson deep in extremely unfriendly Luftwaffe-controlled airspace.

Col. Randal said, “See any problems?”

“It’s not a straight shot the way it appears at first glance, sir. I need to dodge around a couple of islands en route to avoid the possibility of taking antiaircraft fire on a flyover. Also, I have to set up for an approach to come into Hydra Port to meet your requirement to make our approach from the sea. And then try not to end up in downtown Athens when we roll out on the exfil. Fairly straightforward navigation. Easier than locating supply caches in the desert.”

“Questions?”

“Negative, sir.”

Lt. Gen. McKoy said, “Good. Because we ain’t answerin’ any. Get ’er done, Murph. Loose lips sink ships. Don’t let ’em sink yours—it gets cold in Russia.”

“I hear you loud and clear, sir.”

Not satisfied with the answer, Col. Randal said, “I realize I’m putting you in a difficult position by ordering you not to discuss this with Beverly. But that’s the way it has to be, Corporal—I need a Wilco.”

“Wilco, sir.”

Wilco meant, “I understand and *will comply.*” It left no wiggle room for interpretation on the part of the responding party.

Flanigan knocked on the door. Doctor Layton Winthrop arrived. He came in as Cpl. Murphy was leaving. A longtime intelligence officer working in the Aegean/Adriatic area, Dr. Winthrop did not need to be advised about the classified nature of his visit, having already been instructed to be discreet. The professor knew what that implied.

Col. Randal said, “Thanks for coming so quickly, Doctor. What can you tell us about Hydra Island?”

Dr. Winthrop said, “Other than it being named after a mythical nine-headed snake, one of which was immortal, I do not have much on the island. Hercules was tasked with killing the Hydra—an ambitious assignment. At one time the island was a trading post for pirates who plied the Aegean in days past. Some not so ancient, actually. Only the name pirate has changed. Nowadays the freebooters are called smugglers. The island is known for the large mansions owned by the wealthy local 'traders' who ran, and some say still do run, fencing enterprises between the pirates/smugglers and merchants in Greece due to Hydra's immediate proximity to the mainland.

"Reports indicate that following the brutal three-year Italian occupation, most of the mansions and other buildings are in ruins. There are very few men remaining in residence, most having joined the Greek Army, relocated to friendlier climes or been murdered by the Italians. The population is estimated at less than two hundred—the majority being female waiting for their husbands to come home. I have no hard numbers on that."

Col. Randal said, "How would you describe the terrain?"

Dr. Winthrop said, "While I have never visited personally, it is my understanding Hydra is a rugged volcanic island."

Lt. Gen. McKoy said, "Hypothetically speakin' if you were going to conduct a parachute assault on Hydra where would you want to land?"

"If it were me," Dr. Winthrop said, "I would go in by sea. There are numerous small hidden coves to choose from. You could use one of them to make a covered overland approach to your objective. I do not have any information that would lead me to believe the island's terrain is conducive to establishing a proper drop zone without advanced on ground reconnaissance."

Col. Randal said, "Thanks, Doctor. We appreciate your input."

"You are well aware I have my own intelligence sources," Dr. Winthrop said. "Jack is my friend. Whatever it is you are planning, good luck and godspeed. If I did not have the Sikinos commitment I would demand to go with you."

Lt. Gen. McKoy said, "What makes you think we're plannin' . . ."

Dr. Winthrop said, "You two are the best in the business at arriving unexpected and unannounced in the dark of night, kicking down doors and shooting everyone inside. In the event I ever find myself held captive, you are who I want coming to the rescue—covert intelligence gathering, not so much."

After the professor departed, Lt. Gen. McKoy said, "What's he talkin' about? I thought we were pretty good at acquisitionin' target information—clandestine like."

Col. Randal said, "So did I."

LIEUTENANT GENERAL "GERONIMO" JOE MCKOY WAS briefing Colonel John Randal on the work up of Warning Order he had developed to issue to the OSS OG Det 3 Operational Swimmers and Raiding Forces personnel slated for the mission to Sikinos when Doctor Stephen Milam arrived. The surgeon was a recent addition to Raiding Forces. He had come out to the Middle East before the war as a volunteer Royal Army Medical Corps (RAMC) surgeon when the Life Guards and the Blues Cavalry Regiments deployed. While there was no set length established for an overseas tour in the British Army, it was long past time for him to be eligible to rotate back to the United Kingdom. The prewar civilian surgeon to the titled, rich and famous upper-upper-class Six Hundred families, he had elected to extend his tour provided he be assigned to Raiding Forces Headquarters on Castelrozzo as the Medical Officer (MO).

After the last five years, most of it spent as chief of surgery at Cairo General Hospital, he viewed his current duty as a paid vacation on a beautiful Aegean island. Dr. Milam enjoyed the change of pace from the never-ending stress of being the chief of surgery of a wartime surgical unit. He found the assemblage of colorful personalities in Raiding Forces wildly entertaining. He enjoyed the role of being a part of an organization for a change, rather than having to be in charge at all times.

Col. Randal leapt at the opportunity to have such a brilliant, overqualified surgeon on staff for his troops. Major the Lady Jane Seaborn viewed Dr. Milam as part of her family. He had treated the

Fairweather (Lady Jane's family name), Seaborn and Ransom clans for years.

Col. Randal said, "Give me a report, Doctor."

Dr. Milam said, "A 100 percent casualty rate. At least that is what is being reported by the OSS personnel. Possibly a record, one would imagine."

Col. Randal said, "Didn't seem that bad on the DZ following the jump."

Lt. Gen. McKoy said, "Most of the Frogs made it to ABCHQ under their own power."

Dr. Milam said, "The parachute drop was one of those death-defying life-altering events. If an individual was not actually physically injured, he feels compelled to say he was because his friends are all claiming they were. Jump stories galore in the Aid Station. Every Frog was staking out his claim to be the most scared individual in WWII when onboard the jump aircraft en route to Castelrozzo. My medics were laughing so hard they could hardly perform their duties."

Lt. Gen. McKoy said, "Give us some numbers, Doc."

Dr. Milam said, "Out of twenty Det 3 jumpers . . . one fractured radius. Two will require at minimum a couple of weeks of light duty—possibly longer, three more who should be sidelined during any rigorous training for at least a week. Of the rest there are stitches, smaller cuts, bruises and all manner of scrapes and abrasions but the men are cleared to go back to learning how to jump out of an airplane if you deem it absolutely necessary."

Col. Randal said, "Ensign Slade?"

"Westly requested a full body cast, however, he is perfectly fine if a little scratched up. I suspect he only wanted a cast to have the girls sign it."

Col. Randal said, "All right then, put the most serious on profile..."

Noting the confused look on the doctor's face he paused. "Profile" was a U.S. Army medical term. It specified what duty a sick or injured soldier could or could not perform—creating a profile.

"Do what you need to do for the three most badly injured. Provide me the names of the light-duty men. Be advised, if a Frog can walk, I'm going to need him. I realize this is a hard call for a physician to make," Col. Randal said. "If it will make you feel any better, put a notation in the medical records of any individual you have concerns about.

"But keep in mind I have a couple of late-breaking, time-sensitive crises that have to be resolved and no troops to do it with. The exigency of the situation trumps medical prudence or Code of Ethics."

Lt. Gen. McKoy said, "Consider it like a reverse triage."

Dr. Milam said, "I can do that. Out of the original twenty men who made the jump, you have fifteen reasonably fit to fulfill their military duties provided they are not exceptionally arduous. Of the remaining five men, three are borderline marginal to return to duty, two barely able to walk."

"Fifteen walking wounded plus three marginal and two no-goes—got it," Lt. Gen. McKoy said. "Hold the marginals at the Aid Station. I'll be down to inspect 'em in a minute."

Col. Randal said, "I expect full doctor-patient confidentiality on everything we just discussed. And everything we're getting ready to talk about next."

Dr. Milam said, "It is generally considered good form to express that expectation of privilege prior to having the conversation, Colonel. Doctor-patient confidentiality is only applicable as a result of a professional medical relationship. The single time I treated you was the occasion you and Brandy cauterized a through-and-through gunshot in your upper chest with a cigar the way you had seen it done in a Western cowboy movie.

"All I did was sprinkle a little sulfur on the entrance and exit wounds and tape them up with a couple of band-aids. Any Boy Scout with the first aid merit badge could have done as much. Medical treatment based on a scene from a Hollywood flick hardly conforms to the definition of 'best practice.'"

Both of them knew there was no such thing as doctor-patient confidentiality in the military in time of war.

Col. Randal said, "Well, it'll have to be good enough, Doctor, since we never had this conversation anyway. Here's what we need from our MO and I want you to make a production out of it . . ."

As Dr. Milam left the suite he was confident he had made the right decision to request assignment to Raiding Forces. He had not enjoyed himself as much in years. They did not teach this type of practice in medical school. Col. Randal walked him to the door.

Lieutenant Chase Starrett was outside talking to Flanigan.

Col. Randal said, “Come with me, Lieutenant, I’ve been needing a word with you.”

“Yes, sir.”

“Flanigan, do you know where Mrs. Paige can be located?”

“Walked past here on the way to her suite not more than five minutes ago, Colonel.”

“Inform her General McKoy’s on the way to her quarters to brief her about a pair of MI-9 operations.”

“Sir!”

Back inside his office, Col. Randal said, “What were you waiting to see me about?”

Lt. Starrett said, “Sir, six people on the list I drew up for you departed the island before I could notify them they were on Stand-By Ready Alert. I’m sorry, Colonel. It’s the instructors running the SBS officer selection. They’re at sea conducting training raids on non-inhabited keys.”

Col. Randal said, “Never apologize to me, Chase—explain. If I don’t buy your story, I’ll make that apparent. I guess recalling ’em is out of the question?”

“Yes, sir. The training objectives aren’t local. The idea was to give the officers being evaluated time at sea aboard a caique in a leadership position—they sailed hours ago.”

Lt. Gen. McKoy said, “Bein’ down six PAX is goin’ to put a main strain on team configuration for two missions, John.”

Col. Randal said, “No problem. We’ll add The Great Teddy to the list. This mission is all hands on deck. He and Lt. Starrett are my contingency plan. You two can handle replacing six men, right Lieutenant?”

“If you say so, sir,” Lt. Starrett said. He was enjoying working with the Colonel. The assignment had been intimidating at first. He realized he was being evaluated . . . but for what? The job was a priceless opportunity to have a behind-the-scenes glimpse of Raiding Forces senior officers navigating through the problem-solving process with a high-risk mission at the end of it. There was a lot to be learned and he was taking it in.

Something caught Lt. Starrett's attention. Col. Randal mentioned two MI-9 missions. He only knew of the one.

Lt. Gen. McKoy said, "Waldo wasn't on the list either. He'll want in. Man complains a lot but try leaving him out."

Col. Randal said, "Pencil him in, Lieutenant."

"Yes, sir."

Col. Randal said, "Veronica is standing by in her suite. Take Lieutenant Starrett with you and go brief her on our Sikinos mission, General. She doesn't know about Hydra. You might find out if MI-9 has any contacts on the island that might be useful—without mentioning Jack."

Having worked together for years in so many different combat environments, Lt. Gen. McKoy and Col. Randal did not find it necessary to voice every detail about what was or was not expected of each other. When it came to the mission preparation process they knew what the other was thinking. Differences of opinion were rare. There were certain aspects of the operation at this stage neither Mrs. Paige nor Lt. Starrett possessed the Need to Know. They would be filled in later. A few details would only come out after the operation was concluded. In the case of Capt. Jaxx being MIA, secrecy or the impact on morale had nothing to do with loose lips sinking ships. The more important aspect at this point was Col. Randal being prevented from being on the Hudson flying OSS OG Det 3 to the High Speed Launch.

Lt. Gen. McKoy said, "I'll give you a report as soon as I inspect the Frogs."

Col. Randal said, "Take Lieutenant Starrett with you when you do. I want him wired in on the men's physical condition."

Lt. Starrett wondered but was not about to ask what it might be Mrs. Paige knew or did not know about Hydra or why General McKoy was not supposed to mention Capt. Jaxx. But he was curious. It did not escape him that Col. Randal had started referring to the OSS OG Detachment 3 personnel as "the men"—which could be an indicator he was starting to view the Frogs as *his* troops.

Interesting.

COLONEL JOHN RANDAL MADE HIS WAY DOWN TO THE Tactical Operations Center. He saw Major the Lady Jane Seaborn across the room talking to Captain Stephanie Fawcett-Tatum. She waved at him, flashing one of her best-grade smiles. Maybe that meant she was not still mad at him about the Frogs. He was surprised at how relieved he felt about that.

Beverly Blackwell was talking to Wing Commander Paddy Wilcox, DSO, OBE, MC, DFC. He was wearing his trademark black eyepatch covering up a perfectly good eye in order to "strengthen the muscles" in the other one which was also perfectly good. He rotated them. Col. Randal walked over to join the conversation.

Due to tonight's flying missions, the link-up with the HSL and the link-up with LRDG Beach Watch Baker Team 6—being classified as "hasty" based on the short time to launch, limited intelligence and changing mission requirements as new intel came in—the situation was what the military likes to describe as "fluid."

That was about as good a description as any for what was taking place among mission planners in the TOC. It was not very reassuring to Special Operations veterans who pride themselves on preparing for every mission the exact same way every time—Warning Order, detailed planning, Operations Order, rehearsal, drawing specialized mission-specific equipment, test-firing weapons, making radio checks, etc. Typically every man going on an operation knew what equipment every other man carried and what his responsibilities were. If someone went down someone else could step in, take his place and continue the mission. The present situation was not typical. That in-depth mission prep was not going to be possible.

Realistically there is often not enough information or time available to prepare for every operation the way it should be. Raiding Forces trained for those times. Time constraints preventing doing things by the book was not going to throw anyone off their pace.

Tonight's missions, one the staff was aware of and one they were not, were what could be classified as MI-9 rescues—or more accurately extractions—it not being known if either the pilot or Capt. Jaxx had been captured. The ops were not anything like what OSS OG Det 3 had trained for. Nothing had been from the moment the Frogs arrived at Raiding Forces. Since the highly specialized Maritime Unit Operational Swimmers did not know the difference, they accepted that what was taking place was business as usual. They were becoming adaptable, getting used to performing new tasks and beginning to get into the operational rhythm of their new outfit. The Frogs, who thought of themselves as an amphibious quick response unit that could be plugged in wherever the Marines had another beach to land on, had to admit new assignments had been coming at them harder and faster in Raiding Forces than they had anticipated in the Pacific.

Not the duty they had imagined but challenging—the Frogs were up for challenges.

Col. Randal said, "Beverly, we dropped the requirement for you to fly Doctor Winthrop to LRDG Baker Team 6. The Wing Commander's taking that one."

Beverly said, "That's what Paddy was telling me. Excellent, one aircraft handling both assignments simply wasn't going to work. Not enough flying hours to go around."

Col. Randal said, "You said that from the start—why we brought the Commander in."

Wg.Cdr. Wilcox said, "Glad to be of service."

Col. Randal said, "I'll meet the two of you in Doctor Winthrop's office in five minutes. I've got a couple of details to clear up, then we can have what is going to pass as a combined Warning Order/ Operations Order. Wing Commander, you're going to need to take off immediately following the order."

Lieutenant Chase Starrett walked in, spotted Col. Randal and came straight over. “The General’s concluded his inspection—we’re good to go at fifteen Frogs all up, sir.”

Col. Randal said, “You concur with that number, Lieutenant?”

Lt. Starrett had never anticipated being asked if he agreed with a lieutenant general’s assessment of . . . anything. “Those OSS operators are hardcore, sir. Give ’em an aspirin, they’re ready to rock.”

A good answer—just not to the question Col. Randal asked. He let it slide. “Go find the General and Ensign Slade. Have them report to me in the TOC anytime in the next three minutes. You be here too.”

“Yes, sir.”

Col. Randal walked over to speak with Capt. Fawcett-Tatum. “Are those people I asked you to locate in the building?”

Capt. Fawcett-Tatum said, “The Lovat Scouts, Private Hansen and the only two Life Boat Service Men on ABC are in the Ready Room gearing up.”

Col. Randal said, “Send someone to bring them to the TOC for a briefing.”

Lady Jane said, “Mind if I sit in?”

In fact, he did.

Col. Randal had been hoping to get the first phase of the night ahead briefed before she knew it was taking place. Lady Jane did not possess the Need to Know what he intended to do for good reason. The last thing he wanted was for her to put the pieces of the puzzle together and figure out what he was *not* briefing. That could be a problem.

“Absolutely—my pleasure,” he lied.

Lieutenant General “Geronimo” Joe McKoy and Ensign Westly Slade arrived with Lt. Starrett and Lieutenant Ted Hamilton aka “The Great Teddy,” OBE. Col. Randal pulled the general aside for a brief last-minute conversation. One of the Royal Marines escorted the Lovat Scouts, Private First Class Norvel “Horn Dog” Hansen and the Life Boat Service Men into the TOC.

Doctor Layton Winthrop came out of his office having changed into sterile battledress with no badges or insignia. He was wearing a 1911 Colt .38 Super in a chest holster. Lt. Gen. McKoy had given him the handgun based on the lightly customized pistols he and Col. Randal carried. All moving parts were hand polished with special attention given to the trigger and simple higher profile sights than John Browning had designed—gold bead on the front post. The doctor was an excellent marksman. The group moved to the giant wall map where W/Cdr Wilcox and Beverly were talking to Vice Admiral Sir Randolph "Razor" Ransom.

Capt. Fawcett-Tatum had arranged a chair for Col. Randal because of his injured leg. He protested. Possibly a little too much.

Lady Jane ordered, "Take your seat, Colonel."

No one present had heard her use that tone before. Great. His charade was off to a good start.

Lt. Gen. McKoy said, "Colonel Randal asked me to stand in for him for this briefing, bein' crippled and all. Normally he'd be doin' this. Today's Warning Order'll be highly informal and won't follow our normal format because time is short and we don't have much in the way a' actionable intelligence to go over.

"A photoreconnaissance Spitfire outta' the North Africa Photo Reconnaissance Wing commanded by Colonel Elliott Roosevelt was observed by a six-man LRDG Beach Watch team call sign Baker 6 goin' down over Sikinos Island. The pilot was seen to eject from the aircraft and the parachute deployed. The Wing's Commander, Colonel Roosevelt, is President Roosevelt's favorite son. It has been said the Colonel has a taste for five-star hotels, limousines and high-dollar hookers—not necessarily in that order. Used to like to party with Howard Hughes and his contract starlets out in Hollywood.

"Prime Minister Churchill has specifically requested Raiding Forces send a team to recover the NAPRW pilot due in large part to wanting to score brownie points with his counterpart in the White House. General Donovan has confirmed a personal interest in this mission. Lot a' big-time high-profile eyes on this one, boys—be advised this one's political."

The information caused a stir in the room. The Frogs perked up. They were not interested in politics. The idea of a high-profile mission—that was something else.

Lt. Gen. McKoy said, “Professor Winthrop has an agent on Sikinos. Like I mentioned, LRDG Beach Watch Baker Team 6 is located on an uninhabited spit two miles offshore. Immediately upon conclusion of this briefing, a team consistin’ of Dr. Winthrop, Lovat Scout Ferguson, Lovat Scout Fenwick, Life Boat Service Man Bleecker, Life Boat Service Man Greathead and Private First Class Hansen actin’ in the role a’ team radio operator, will board a Supermarine Walrus aircraft piloted by Wing Commander Paddy Wilcox. They’ll be flying up the coast a’ Turkey to the Levant Schooner Flotilla schooner moored closest to Sikinos—LSF Schooner 7. Mrs. Brandy Seaborn’ll link up with the professor’s team when they arrive and she’ll transport it to Baker Team 6 by MAS boat.

“From there a decision’ll be made whether to load up the LRDG and take everyone over to Sikinos to rendezvous with the professor’s agent on the MAS boat or to have him, the Lovat Scouts and Horn Dog paddled over in the Baker Team 6 rubber assault raft. Either way, based on information the professor’s man provides about the pilot’s current location, a plan of action’ll be drawn up and a’ extraction initiated. Intel reports indicate you can expect a squad a’ 999th Light Division bad guys to be stationed on the island pullin’ occupation duty.

“Simultaneous with the professor’s team’s movement to Sikinos, the Hudson, flown by Beverly Blackwell, will be headed toward a Small Raids Inc. High Speed Launch located at sea seventy-five miles off the island. Once there, a fifteen-man team of OSS Operational Swimmers plus attachments under my command’ll disembark the Hudson and load on the HSL. The boat will then make a high-speed run to Sikinos and stand offshore awaiting developments—ready to land the troops if so requested.

“Wing Commander Wilcox’ll return to LSF Sloop 7 in the Walrus and stand by in the event his services are needed on the exfil. Mrs. Seaborn’ll remain offshore Sikinos in the vicinity of the LRDG Beach Watch location prepared to extract the professor’s team if called on. After dropping off the Frogs Beverly’ll return to Castelrozzo.

"Doctor Winthrop will make contact with his agent. Establish where the pilot is located. And then make a determination on the next course of action.

"This concludes my briefing. What are your questions?"

The few long-service veterans making up Dr. Winthrop's team recognized they would be winging it on this mission. What Lt. Gen. McKoy had laid out was a complex scheme of maneuver designed without the benefit of hard intel, advance reconnaissance or face-to-face coordination between all of the key players. The Frogs did not ask questions. Not because they did not have any but because they feared asking questions might show their lack of experience.

Lt. Gen. McKoy said, "Once you load out stay loose boys. This thing's fluid—there'll be change mission frag orders in your future, guaranteed."

Col. Randal realized he had a problem.

As soon as Lt. Gen. McKoy gave the traditional "This concludes my briefing", he said, "General, I need a word."

The two stepped off to a corner as the professor and his team prepared to move to the dock to board W/Cdr Wilcox's Walrus.

"Dr. Winthrop, stand fast. I need to speak to you as well as soon as I have a word with the General."

Lt. Gen. McKoy said, "What's up, John?"

Col. Randal said, "The Sikinos phase isn't going to work the way we've set it up. When you pull out of the Frog team transferring to the HSL at the last minute, Ensign Slade's next in the chain of command. He isn't experienced enough for this mission. There are no other officers available except Lieutenant Starrett. He's fully capable but putting him in over Ens. Slade at this point could cause resentment among the Frogs."

"What do you have in mind, John?"

"I need you on that HSL," Col. Randal said. "The minute it arrives offshore at Sikinos, take charge of the entire operation—Dr. Winthrop's team, the LRDG Baker Team 6 and OSS Det 3. Evaluate the situation, develop a scheme of maneuver based on what you see once you're on the ground, then execute it. You'll have to coordinate by radio but no one will question your authority or right to command."

Lt. Gen. McKoy said, "My pleasure, John."

Col. Randal turned toward the professor. "OK, Doctor now here's what . . ."

If there was ever a textbook example of how *not* to conduct a long-range air, sea, land Combined Operation, this was it.

CAPTAIN BILLY JACK JAXX WAS SITTING ON THE SECOND-floor verandah of the police station gazing out over the bay talking to the chief of police. It was beginning to be late in the day. The sun was softening. They were drinking traditional Greek espresso out of a glass. It was named either Turkish coffee or Arab coffee. Neither one made much sense but that was what the policeman said. Whatever it was called, the drink packed a punch.

Capt. Jaxx was enjoying himself. If he had one of Waldo's cigars to smoke, the approaching twilight and good company would have been the almost perfect end to a long day.

The chief said, "What is your plan, Captain Billy?"

Capt. Jaxx said, "I transmitted a distress call on the Germans radio. My people will come pick me up. No problem."

The chief said, "You should hope they arrive first."

"First?"

"The Germans are scheduled to relieve the detachment on our island today. Once a month a fresh squad from the 1st Mountain Division rotates the duty—usually it's here by now. The Nazis are not going to be pleased you killed their men."

Capt. Jaxx said, "You're kidding."

7

PATELLOFEMORAL PAIN IN THE PATELLA

DOCTOR CLAYTON WINTHROP DEPARTED TO BOARD THE Walrus flown by Wing Commander Paddy Wilcox. There was so much going on in the Tactical Operations Center that the traditional escort to the dock was not on for today. The professor was relieved to learn he would be handing off command of the rescue mission to Lieutenant General "Geronimo" Joe McKoy. He much preferred the role of spymaster working in the shadows. The professor was accompanied by the Lovat Scouts—who had been issued strict orders not to let anything happen to him, Life Boat Service Man Stanley Bleecker, Lifeboat Service Man Henry Greathead and Private First Class Norvel "Horn Dog" Hansen, the team radio operator. They simply walked down to the dock and boarded the plane. No fanfare. Professionals commuting to work.

In the TOC the fifteen members of OSS OG Det 3 classified as fit for duty and confirmed as such by Lt. Gen. McKoy were sitting in chairs assembled at the giant wall map of the Aegean Theatre of Operations. James "Baldie" Taylor arrived, having been notified by Major the Lady Jane Seaborn that he was invited to travel to London with her party—knowing he needed to be along to advise the Chief of the Office of

Strategic Services, Brigadier General William "Wild Bill" Donovan on intelligence matters. He went into conference with Vice Admiral Sir Randolph "Razor" Ransom and Brigadier Raymond J. Maunsell, who liked to be called R. J. Aware of the mission to bring out the Northwest Africa Photo Reconnaissance Wing pilot he had flown in from Cairo with Jim to see if he could be of any assistance. Lady Jane, Beverly Blackwell, Mandy Paige and her mother, Veronica, walked in and took their places in chairs provided for them. Waldo Treywick came over and took a seat with the OSS men. They were surprised to discover he was going tonight. The former ivory poacher was armed with Colonel John Randal's old cut-down Browning A-2 12-gauge semiautomatic shotgun which was his weapon of choice for fast close-in night work. His selection of weapons, there was also a big Triple Lock Smith &Wesson 38-40 double action revolver in a shoulder holster on the outside of his M-42 jump jacket, was a clear indication of how seriously he was taking what lay ahead. Waldo's unexpected arrival in their midst caused a stir among the Operational Swimmers. The man had a reputation.

When Col. Randal arrived—noticeably limping—Lieutenant Chase Starrett, a product of the Virginia Military Institute's accelerated wartime commissioning program—meaning he had been commissioned after his sophomore year—instinctively called, "ATTENTION!"

He realized he had made a mistake the instant he did it. The custom was not normally practiced in Raiding Forces. And he knew that.

Everyone jumped to their feet.

Col. Randal ordered, "At ease—as you were."

When everyone, except those who outranked him and had not come to their feet, had retaken their seats he limped to the front of the group. The number of onlookers had grown considerably as everyone in the TOC drifted over to listen. All business, Col. Randal cut straight to the chase. "Situation. . ."

Doctor Stephen Milam and two of his medical orderlies burst through the door of the TOC moving fast. One was pushing a wheelchair. Dr. Milam, who held the military rank of Brigadier in the Royal Army Medical Corps but never used it, preferring to be called Doctor, ordered in a stern command voice, "Stand down, Colonel!"

Everyone froze. One sacred rule not to be violated in Raiding Forces was "an officer in the act of issuing an order to troops was never interrupted"—ever. All questions were saved until he had completed his briefing, the idea being not to interrupt his train of thought. Then the briefer would take questions until they were all satisfactorily answered. There was no limit and people could ask more than one provided they were not argumentative.

Dr. Milam said, "Colonel, you failed to comply with my explicit instructions to report to the Aid Station for a medical review. If you refuse to come to us we shall come to you. Hop up on that table there."

There just so happened to be a table conveniently stationed next to the wall map where everyone in attendance could see it.

"Doctor, can't you. . ."

"Your options, Colonel, are to get on the table and allow me to examine your leg or I shall have you restricted to hospital for a minimum of twenty-four hours for a more complete evaluation—take your choice."

Col. Randal was visibly angry but he hoisted himself up on the edge of the table. Dr. Milam and the two orderlies gathered around. A careful inspection was made of his leg. The doctor took his time dictating his findings to one of the medical orderlies who wrote them down in shorthand on a notepad.

Col. Randal was shocked by his injured leg's lackluster response when Dr. Milam tapped his knee with a small triangular hard rubber neuro hammer. Normally doctors best be advised to stand off to one side when they performed that test. He had excellent reflexes.

"What's the verdict?"

"Patellofemoral pain in the patella."

"What does that mean?"

"Severe knee strain. And for you it means light duty for a minimum of six weeks. Possibly longer."

"Are you crazy?"

"It could take months to fully recover, Colonel."

Col. Randal glanced over and made eye contact with Lady Jane on the front row. She was dialed in listening to every word—no heart attack or any other kind of smile. "What's the rehabilitation path look like?"

"Rest, walking, swimming, yoga, water exercises, ice. . . more rest.

"A joke, right?"

"Massage is helpful. You have your slave girls, Colonel. Crack the whip, put the two of them to work. They should be of more use to you than dancing at the Kit-Kat."

When Col. Randal made a move to slide off the table Dr. Milam said, "Not quite yet. One last detail. Tape him up, gentlemen."

The two orderlies were standing by waiting for those instructions. They moved in with a purpose. First they rotated Col. Randal until his injured leg was lying stretched out flat on the table. Splints were slapped on from his crotch to his ankle and the entire leg was taped from top to bottom. When attempting to stand he nearly toppled over. Crutches were handed to him. A new life experience.

Col. Randal said, "So how am I supposed to do the walking, swimming, water calisthenics etc. with this contraption strapped on my leg?"

Dr. Milam said, "Hopefully we shall be able to remove the splint in six to ten days. Ice, massage and limited walking can commence immediately. No restrictions within the parameters I have laid out, but do not overdo it. What your knee requires to heal properly is time."

Col. Randal said, "Lovely—can I ride in a car or a plane?"

"Absolutely, but I would recommend against attempting to drive."

Standing up awkwardly supported by his crutches, Col. Randal picked back up where he had been interrupted. "Situation. . ."

The fifteen Office of Strategic Services Operational Group Detachment 3 Operational Swimmers medically cleared for the night's mission watched in awe. This was one for the books. One team had departed to be flown to the objective by a one-eyed pilot wearing an eye patch and now a wobbly full colonel on crutches was giving a Five Paragraph Operations Order for a mission with national-level implications as if it was business as usual. Nothing to see here. Unbelievable.

Until arriving at Castelrozzo the Frogs had been living the life of military nomads traveling from one training site to the next, always on temporary duty (TDY). Now at long last they found themselves at their permanent duty station assigned to a hard-charging unit commanded by a CO who did not allow being crippled to slow him down. Raiding Forces was a serious outfit.

Col. Randal said, "This concludes my briefing."

The Frogs surprised everyone in the TOC when they stood up and applauded. Col. Randal was clearly hurt but pressing on, setting the example. They respected that.

OSS OG Det 3 was fired up about the opportunity to rescue a pilot from a USAAF wing commanded by one of the President's sons—a glory mission. No need for questions. They knew who, what, when, where . . . the how, not so clear.

The Operational Swimmers were eager to get boots on the ground and make it happen. The men were in lockstep agreement that their new commanding officer had issued one of the best Operations Orders they had ever heard. And that included the canned, much-rehearsed presentations by professional instructors at the best schools the U.S. military system had to offer.

What the Frogmen did not know . . . it was almost all fiction.

BEVERLY BLACKWELL AND MANDY PAIGE RUSHED UP TO Colonel John Randal as soon as the briefing was concluded. The girls were worried. Both hugged him—which was not done in the TOC during duty hours and since the TOC was always running, that meant never.

Mandy said, "I had no idea you were so seriously injured."

Beverly said, "Does getting bucked off your saddle animal qualify for a Purple Heart?"

Lady Jane came over. "Volunteer masseuses no doubt."

Col. Randal said, "You put Doctor Milam up to that?"

Lady Jane laughed. "No, but I wish I had."

The Frogs were heading to the Ready Room to collect their gear in anticipation of moving directly to the dock to load out on the Hudson. Col. Randal hobbled over and pulled Ensign Westly Slade aside. No one was within hearing distance.

Keeping his eyes sweeping the TOC to make sure no one was approaching, Col. Randal said, "We are not having this conversation."

Ens. Slade had been waiting for someone to say something like this to him ever since he volunteered for the Office of Strategic Services. "I hear you Love and Charlie, sir."

Col. Randal said, "Prior to the link-up with the High Speed Launch you'll be issued a change of mission. General McKoy's taking command of the overall Sikinos operation at that time. Be advised that once your Frogs land ashore OSS OG Det 3 personnel are likely to engage in intensive fighting. It might be a good idea to reevaluate your men's individual weapons choices and the basic load of ammunition they carry."

Ens. Slade said, "We brought Model 42 Marlin 9mm submachine guns that have not been unpacked yet. During testing at Aberdeen at one hundred yards they outperformed the Thompson and the Sten Mark II—they're real scorchers. Will those work, sir?"

Col. Randal said, "I've never heard of the Model 42 Marlin. If you believe they'll increase your team's firepower, take 'em. It's your call, Ensign. Time to do some small unit leadership stuff."

"We'll break out the Marlins, sir," Ens. Slade said. "I'll have my people carry a double basic load of ammo. Extra demo."

Col. Randal said, "Make it fast—see you at the Hudson."

Lieutenant Chase Starrett came straight over when Col. Randal made eye contact with him across the room. They waited until Ens. Slade was gone. "This conversation does not go any farther than the two of us standing right here, Lieutenant."

"Yes, sir."

"I need a Wilco."

"Wilco."

Col. Randal said, "Should the subject come up at a later date it's in your best interest to stick to the plausible story it never took place—I'll back you up."

Lt. Starrett said, "I can do that, sir. I'm pretty good at stupid."

Col. Randal said, "No one has the Need to Know what I'm about to tell you. Particularly Lady Jane, her Royal Marines and/or any of the other female personnel on ABC to include Greek island women."

"Understood, sir."

Col. Randal said, "Captain Jaxx's submarine, the *Perseus,* has gone down with all hands. Our Y-Service radio intercept people picked up a garbled transmission coming from Hydra Island claiming to be from him. The Admiral says he doesn't believe it's a legitimate signal. He suspects a trap.

"Prior to the Hudson linking up with the HSL Ensign Slade will be issued a change of mission. Only Waldo, five Frogs and the Ensign will disembark. They'll be led by General McKoy who'll be assuming overall command of the Sikinos mission. As soon as the Hudson takes off with the remainder of the Det 3 team still onboard Beverly will be provided a new flight plan. You'll be the acting senior officer at that point."

Lt. Starrett said, "Why all the secret squirrel stuff, sir?"

Col. Randal said, "I'm planning to be on board the Hudson. If Lady Jane should discover what's happening, she'll probably find some way to have me confined to quarters to prevent me from going. If Beverly finds out, I wouldn't put it past her to invent some excuse not to fly the second leg of the operation. More importantly, if Admiral Ransom hears any of this, he might order me not to travel out with you men."

Lt. Starrett said, "You have something that crazy planned, sir?" He was wondering about the "second leg" of the flight plan.

Col. Randal said, "Violates every one of our Rules for Raiding and most of the unwritten stuff. No Plan B. Because we don't have a Plan A."

"What could possibly motivate you to break every one of our Standing Orders, sir?"

Col. Randal said, "That's classified."

Dr. Milam was waiting for a word with Col. Randal when he finished his conversation with Lt. Starrett. “How do you feel our little soap opera went?”

Col. Randal said, “Overkill with the splints, Doctor. That wasn’t part of the script. The wheelchair was a nice touch.”

Dr. Milam laughed. “I took an elective drama class at King’s. Nearly destroyed my First Class Honors standings due to my love of improvisation and overacting. Should have warned you about that, actually.”

“Yeah.”

“Be advised, Colonel, you do have a knee injury. Not anything as serious as I described but pamper it or there shall be consequences,” Dr. Milam said. “No additional trauma. Yoga classes with Happy—I hear he is quite the performer—slow walks, etc. Ice and massage are therapeutic, swimming, very good.”

Col. Randal said, “Roger, I’ll keep that in mind.”

Which was not the same as “Wilco.”

Dr. Milam handed Col. Randal a soft knee brace, “Keep this in your pocket. You might find need for it. In case the urge to do something stupid overwhelms your normally first-rate sense of judgment.”

“Thanks, Doctor.”

Lieutenant Ted Hamilton aka “The Great Teddy” walked past heading to the Ready Room to gear up with the Frogs. Col. Randal pulled him aside. “Here, take my handgun, Lieutenant. Go up to my suite. Have Flanigan let you in. My other pistol and my suppressed .22 High Standard are on the coffee table. Stick both Colts in the pack containing your 45mm rounds along with six spare magazines of 38 Super. Don’t let anyone know you’re doing it.

“Bring me the .22 and my chest holster—be discreet.”

Lt. Hamilton said, “Sir!”

“Tell Flanigan I said he never saw you. You were never there.”

None of this made sense to Lt. Hamilton but that was likely the idea. He tried to not ask questions when Col. Randal gave him a direct order, “Roger sir—hey, presto!”

Col. Randal said, “That’s the spirit.”

COLONEL JOHN RANDAL LED OUT FROM ABCHQ ON HIS crutches. OSS OG Det 3 followed in single file behind him led by Ensign Westly Slade at a slow walk. Lieutenant General "Geronimo" Joe McKoy, Waldo Treywick, Lieutenant Chase Starrett and Lieutenant Ted Hamilton brought up the rear. The movement to the dock occurred so quickly that the people in the TOC did not have time to gather to see the team off, which was the idea. With Major the Lady Jane Seaborn leading the way everyone rushed out to catch up.

Col. Randal was having a difficult time navigating the cobblestone path down to the wharf on his crutches. He was cursing Doctor Stephen Milam under his breath. Happy ran up and brushed against his leg. The dog was being friendly. It nearly knocked him off his feet. Going downhill on crutches for the first time was a high-risk on-the-job training proposition.

No drums were beating like in the movies but there should have been. OSS OG Det 3 was going on its first Combined Operation as a unit and they were doing it almost entirely on their own. Not optimal for such a multifaceted highly complex mission. Under normal circumstances, the Frogs should have had experienced men from Sea Squadron, the Lounge Lizards or the 575th Ranger Force embedded as stiffeners but due to time constraints and for reasons of his own, Col. Randal had not chosen to do things that way.

Because Col. Randal was hobbling along on his crutches the column was moving so slowly that Lady Jane and Mandy were able to catch up to him.

Mandy said, "Where is your Colt automatic, John? You practically sleep with one."

Col. Randal lied. "I was afraid the holster would hang up on these crutches."

"Give those to us," Lady Jane said. "We can help. Put your arms around our shoulders."

Col. Randal said, "That's not. . ."

"Just do it, John."

"Yes ma'am."

Mandy said, "Try to be a good patient. That is what you would tell us if we were in your place."

Wanting to change the subject Col. Randal said, "Jane, did you know the Frogs intend to put you in for a medal?"

Lady Jane said, "Whatever for?"

Col. Randal said, "Westly tells me that after the plane was attacked you saved them by leading the stick out of the plane prior to the green light."

Lady Jane laughed. "You always say it does not count when saving oneself. The plane was on fire. I wanted out."

Col. Randal said, "Apparently the decoration's a group decision. You can't decline something like that. A telex has been dispatched to General Donovan describing the action for the citation. These OSS operators don't fool around once they make up their mind to do a thing."

Beverly had the Hudson's engines running when the column arrived at the dock. She was ready to get the show on the road not wanting to run out of daylight. Finding the HSL might not be as easy as Murph the Surf made it sound. Col. Randal and his two helpers led the file down to the landing at a funeral march pace. Then he stood by the door to observe as the Frogs loaded, making eye contact with each man as they came past. The Frogs acknowledged him with nods as they passed. But the truth was they were much more interested in Mandy, who was helping Lady Jane prop him up—they had not yet met her.

Lt. Gen. McKoy, Waldo Treywick and Lt. Starrett were the last to board. After they loaded Col. Randal said, "I'm flying out with Beverly."

Lady Jane said, "Not a chance . . ."

Col. Randal said, "You heard Doctor Milam authorize me to ride in a vehicle or airplane."

Before Lady Jane could stop him he lurched forward, semi-diving headfirst as the Hudson started taxiing.

Lt. Starrett was standing by to help pull him in, "Well played, Colonel."

"Looks like you've got a couple of angry females back there, sir," Ens. Slade said. "Probably better we can't hear what it is they're screaming."

Lt. Gen. McKoy said, "Gone from wantin' to save you to wantin' to murder you, John—wimmen can do that."

Col. Randal said, "Yeah."

Ens. Slade said, "Must be nice having all the ladies you have in your life looking out for your best interest. You don't make it easy for 'em do you, sir?"

Col. Randal said, "Cuts both ways."

Lt. Gen. McKoy said, "Ain't that the truth."

COLONEL JOHN RANDAL HUDDLED WITH LIEUTENANT General "Geronimo" Joe McKoy, Lieutenant Chase Starrett, Lieutenant Ted Hamilton, aka "The Great Teddy," Ensign Westly Slade and Waldo Treywick in the tail of the plane conducting an officer's call. The rear door on the port side was open and they could see the Aegean flashing past almost level with the Hudson. Beverly Blackwell was skimming the waves in order to make it harder for marauding Luftwaffe fighters to spot the plane. On the return trip after nightfall she would pull up to normal cruising altitude.

Col. Randal said, "We have a change of mission. The submarine Captain Jaxx was on went down with all hands—declared 'missing' which in navy-speak for subs means lost. A radio transmission claiming to be from him has been received coming from Hydra Island. The Admiral says there is almost zero probability Jack was the actual sender. He believes there's a high likelihood German Naval Intelligence broadcast the message to lure us into dispatching a rescue team.

"That's exactly what we're going to do."

Lt. Gen. McKoy said, "A cutter from the High Speed Launch we're rendezvousing with will be standing by as soon as we land. I'll disembark with five Frogs and Waldo. Westly, pick the team for me.

Once onboard the HSL I'll assume overall command of the Sikinos operation.

"Ensign Slade you're my deputy team leader. I want you to designate a Frog to be my RTO. I can work the radio, I just need someone to tote it and monitor incoming messages when we're on the move."

Ens. Slade said, "Roger, sir."

Col. Randal said, "Beverly will take off with the remainder of Det 3's personnel still onboard. They'll be flown to Hydra Island to extract Captain Jaxx—the second phase of tonight's operation."

No one said anything but this was not even a complete Frag Order. They were reasonably sure Col. Randal knew that. It was beginning to dawn on Lt. Starrett, as the senior combat effective officer remaining in Det 3 he had just been handed command of a rescue mission with zero intelligence on an island he had never heard of and had no idea where to locate the person needing to be rescued on top of which it was zero dark thirty. The assignment was more technically daunting than leading his stick on a fighting withdrawal of over a hundred miles through enemy territory after the misdrop on Benevento.

Lt. Gen. McKoy said, "Westly, have my team a' Frogs move to the rear of the plane so they can be first off—I want a fast exit. You boys do know how to board a dinghy at sea?"

Any other time that would have drawn a laugh. "Yes, sir."

There was no scenario in which Ensign Slade had ever envisioned a five-man OSS OG Det 3 team being led by a lieutenant general. A strange day had turned stranger. This mission made swimming off an enemy shore in broad daylight blowing up obstacles in advance of an amphibious assault by U.S. Marines on a Jap-held island somewhere in the Pacific sound like a casual day at the beach.

After Ens. Slade went to select Lt. Gen. McKoy's team, Waldo said, "Colonel, I think I'd rather travel with Lieutenant Starrett—go get Jack."

Lt. Gen. McKoy said, "I'll tell Westly to pick us one more Frog to fill your slot."

At this point, Waldo had no inkling he had volunteered to jump on Hydra. He was an unenthusiastic paratrooper at best. Knowing this would not have changed anything. Had their roles been reversed Capt. Jaxx would have come for him.

Normally a flight or transport by boat to reach the target area for a mission seemed to take forever. It was always that way. Not so today. Time was racing by. Col. Randal hobbled over to where the Frog team would be exiting. Lt. Gen. McKoy was sitting on the end of the canvas bench seat. He had the men lean forward to give them a short pep talk. OSS OG Det 3 all had a couple of missions under their belts but that experience only served to demonstrate to them what they needed to be worried about.

The Frogs were wired tight.

Lt. Gen. McKoy kept it short and simple in compliance with Raiding Forces Rules which they were in the process of breaking. "When I say frog, you boys jump."

When the General finished Col. Randal said, for those of you who don't know General McKoy is a Medal of Honor recipient and a veteran of at least five wars I know of. I expect you to give him your best effort one hundred percent."

Then Col. Randal struggled to his feet. He shouted loud enough to be heard by all of the OSS operators onboard. "I could have brought in my veteran Raiding Forces people for this mission but chose to go with you men. Why? Because Det 3 has been working together for the last two years. You know each other's tendencies and capabilities. Those are the basic building blocks of a special operations team purpose designed to be tasked with high-risk high-reward assignments. In Raiding Forces we like to say 'improvise, adapt, overcome.' That's what I expect in a big way tonight. This is your chance to show what you can do.

"Don't let me down."

As the OSS operators were responding with "Hell yesses", "no sirs" and "roger that's"—though the men not now disembarking with Lt. Gen. McKoy were in the dark about what role they were going to play," Beverly came on the intercom.

"HSL in sight. Standby for landing. Let's hear it for Murph the Surf back there, boys—he nailed it."

Cpl. Murphy's performance was extraordinary long-distance aerial navigation over a large body of water—meaning he did not have terrain features to guide by. Hitting the Royal Navy's High-Speed Launch dead-on was a navigator's equivalent of a hole-in-one. The Frogs cheered, but the men on Lt. Gen. McKoy's team preparing to disembark were somewhat distracted by what was immediately ahead for them.

Lt. Starrett said, "I thought Murphy was puffing his navigational skills."

Lt. Gen. McKoy said, "Ain't braggin' if it's true."

The General's team disembarked. As soon as the cutter was away rowing back to the HSL, Col. Randal started to hobble to the cockpit. The Hudson was already beginning its takeoff run. Beverly believed her part of the operation had been completed was heading home and not wasting any time about it.

She had no idea Col. Randal was on board her aircraft.

DOCTOR LAYTON WINTHROP'S TEAM CONSISTING OF LIFE Boat Service Men Stanley Bleecker and Henry Greathead, Lovat Scouts Lionel Fenwick and Munro Ferguson and Private First Class Norvel "Horn Dog" Hansen were aboard a Walrus piloted by Wing Commander Paddy Wilcox, winging their way up the coast of Turkey en route to Levant Schooner Flotilla Sloop 7. Captain Jeffery Tall-Castle, MC, OBE—a long-time Raiding Forces officer with a record of distinguished service with Force N in Abyssinia, as a Pathfinder during the drop on Persia and as a jeep patrol commander out of Oasis X—was onboard the sloop commanding a troop of three teams of caique raiders.

Two of the teams were currently away on independent operations and one was standing down having recently returned from a mission. This was the Constant Pressure Concept in action. Pinprick raids being carried out around the clock throughout the islands in Capt. Tall-Castle's assigned AO. Two teams out raiding and one refitting, rearming and resting. In the future, patrols on standdown would be returning to Castelrozzo for retraining prior to going back out.

LSF 7 was moored twenty nautical miles from the spit the Long Range Desert Group Beach Watch Team 6 was ensconced on two miles off Sikinos. The professor's plan was basic, conforming to Raiding Forces Rule 2, "Keep it Short and Simple." W/Cdr Wilcox would land at the schooner and the team would wait until nightfall. Then a MAS boat skippered by Brandy Seaborn would transport them to Beach Watch Team 6 and stand by for the professor to decide on his next course of action.

There were essentially two options with a number of variations available to him. Brandy could drop Dr. Winthrop's people off and return to LSF 9 and await orders. Or she might stand by at the spit while the professor's people paddled over to Sikinos, linked up with his agent, recovered the pilot and paddled back. Then she could transport Dr. Winthrop's team, LRDG Beach Watch Team 6 and the precious cargo to Advanced Base Castelrozzo by sea. Or the team could exfil aboard the Walrus. There were other possible options for the return but those would be evaluated as the operation unfolded.

There were a lot of variables, a lot of unknowns and a lot of moving parts. Not every mission is clean and neat prior to. While it is best to have current intelligence, plan every detail in advance, develop contingency plans for every possible eventuality and rehearse for days, weeks, or even months, that is not always doable in a fluid, fast-developing combat environment.

Many things could go wrong. There was only one good outcome. Heavy responsibility was being placed on Dr. Winthrop to adapt to the situation on the ground once he went in. He was going to have to take it one step at a time, carefully calculating each move. There was one factor in his favor. Colonel John Randal had provided him with a team composed of the best men he had available.

LIEUTENANT GENERAL "GERONIMO" JOE MCKOY assembled his five OSS Operational Swimmers—six counting Ensign Westly Slade—on the stern of the High Speed Launch as soon as the boat was underway. The HSL was a sixty-four-foot high-speed air/sea rescue launch built by the British Power Boat Company (BPBC). The Type II, like the one they were aboard, was called a "Whaleback" because of the curve to its deck and its unusual humped cabin. While it was designed to go forty mph, both of the Raiding Forces MAS boats were as fast and much more heavily armed.

HSL's standard organic weapons consisted of two .303 Lewis machine guns mounted in single turrets on the roof of the humped cabin for air defense. The launches had been developed for the RAF's Marine Craft Unit before having recently been transferred to Vice Admiral Sir Randolph "Razor" Ransom's Small Raids Inc. It was reasonable to expect Admiral Ransom would be upgrading their firepower in the near future. He had a fetish about that. Depth charges being a simple add-on had already been fitted. Specialist craft have their place but not in a small unit navy war like was being fought in the Aegean. Once heavier armament was retrofitted, the HSLs would be perfect for island raiding while continuing to be able to serve as air/sea rescue.

Lt. Gen. McKoy said, "All right boys, here's the deal. As you know our mission is to rescue President Roosevelt's favorite kid's photoreconnaissance pilot. Listen up, dial in and focus on the task immediately in front of you at all times—don't be daydreamin' about photo ops at the White House. Once we get ashore changes'll be comin' hard and fast outta the blue, so stay loose. Show me that lightnin' quick response time they tell me you Frogs are capable of. Do your job and you'll all be heroes. Just might get your picture in *LIFE* magazine. Who knows? Do wonders for your love life—you can quote me on that."

The Frogs laughed.

"One more thing and it's significant. You'll be goin' against the 999th Light Division. You know who they are. They've got the German military designation as 'Criminals' for a reason. We generally don't take those people prisoner—ain't got time tonight and nobody to spare guarding 'em.

"Anybody got any questions?"

The Frogs were thinking one thought with one mind. "*LIFE* magazine"—how fast can we get ashore and start killing Nazis?"

AS THE SKY WAS TURNING CRIMSON FROM THE SUN starting to go down, Captain Billy Jack Jaxx was stashing 7.62 SVT-38 semiautomatic rifles and Steyr 9mm submachine guns in various fallback positions. He had recovered three more SVTs from the ski trooper's barracks. And now he had six 9mm Steyr MP-34 submachine guns. They were secreted in places that offered a good field of fire all the way down to the pier. He was planning to fight a hit and run battle—fire and fall back on the single street in the village starting twenty-five yards from the dock where the police chief confirmed the replacements always landed.

Capt. Jaxx worked his way up the narrow road to the end of the buildings establishing additional fallback positions as he went. Since the dock was the only one in the village it was reasonable to expect the boat bringing in the 1st Mountain Division's replacement squad to dock where the PC said. The plan was for him to make a fighting withdrawal, leapfrogging up the street from one fallback position to the next. Then hopefully he would disappear into the interior of the island as night arrived, knowing full well *hope* was not a course of action.

It was not much of a plan but it was the only one he had. You fight with what you have. Not what you want.

Right about now Capt. Jaxx was beginning to regret not taking one of the caiques and sailing away over the horizon. However, having grown up in semi-arid West Texas with the nearest lake over a hundred miles away, except for the small boat handling taught at the Commando Basic Training Center in Scotland—which consisted mostly of rowing assault boats of various types ashore with the idea of depositing small raiding parties on an enemy coast to do the King's mischief, he had no idea how to handle a caique. Or to perform celestial navigation at sea there being no terrain features to guide by.

Besides, he did not know where he was other than the police chief having said it was Hydra Island. Wherever that was. So the sailing off over the horizon into the sunset option was still a nonstarter.

Jack Cool was prepared to do battle with the hand he had been dealt. "Let's rock."

COLONEL JOHN RANDAL STRUGGLED TO THE COCKPIT OF the Hudson, the splint on his leg making movement around the aircraft difficult. Beverly Blackwell was in the pilot's left seat. Corporal Tom Murphy aka Murph the Surf was navigating from the co-pilot's seat. When he hobbled in Beverly screamed, "Johnny, what're you doing here?"

Col. Randal handed Cpl. Murphy a sealed envelope. "This is your new flight plan, Corporal—make it happen."

"Sir!"

"Do it from the navigator's station—give us the cabin."

When Cpl. Murphy had cleared the flight deck Beverly said, "Does Lady Jane know you're on this airplane?"

"Yes, she does."

"With your injury she gave you permission to be on this flight?"

"Not exactly."

"How much not exactly."

"Somewhere in the realm of a hundred percent."

"You're in big trouble, cowboy."

"The submarine Jack was on went down with all hands."

"Oh!"

Beverly started sobbing.

Col. Randal said, "A garbled radio signal claiming to be from him was picked up by the Y-Service Signals Intelligence people. They passed the transmission on to the Razor. I've given Murphy a new flight plan to the island—a place called Hydra."

He thought it best to leave out the parts about the Admiral believing it was virtually impossible to escape from a submerged submarine given the current state of the escape equipment currently available. And that the Razor thought the garbled message was a ploy by German Naval Intelligence, Marinenachrichtendienst (MND), to lure a rescue team into an ambush.

Col. Randal said, "By my rough estimate we should arrive off Hydra right at sundown—maybe sooner. When we do, I want you to line up on the single street running up from the pier at 500 feet altitude and cut your speed back to 110 miles per hour."

Tears were streaming down the UT beauty queen's cheeks. "What's going on, Johnny?"

Col. Randal said, "There's ten Frogs in the back with Lieutenant Starrett, Lieutenant Hamilton and Waldo. They'll be chuting up in a minute. You're going to green-light a jump when you fly over the dock—make your approach coming in from the sea.

"Run your engines at War Emergency Power if you need it to get there before dark. I don't care if the Hudson's a write-off after we return to ABC. Following the drop, racetrack until you see a series of flares, red over green. Then land at the pier and we can load up and go home."

Red over green flares were the traditional Commando signal for "mission accomplished."

Beverly said, "You're going to put those no-training, one-jump, non-volunteer paratroopers out over the middle of an enemy-held built-up area in broad daylight—have you lost your mind?"

Col. Randal said, "You have a better idea?"

Beverly said, "This is a crazy, awful, really bad idea, Johnny."

Privately willing to concede it was the worst he ever had but not wanting to admit it out loud, Col. Randal said, "Lighten up, Beverly. We'll rescue Jack."

"You don't even know where he is. How big is Hydra?"

"Approximately twenty square miles."

Beverly said, "General McKoy was right you must have landed on your head when you got pitched off your mule."

Col. Randal staggered back to the troop compartment. The Frogs were sitting on the canvas bench seats no doubt wondering what was about to happen. Standing in the rear of the plane studying the men who were all turned to face him he ordered, "Listen up people. It's Raiding Forces SOP for X-type parachutes to be carried aboard all jump-capable aircraft. You'll find them stowed under the seats you're sitting on. Reach down and secure one for yourself."

This announcement was met by shocked expressions as the OSS OG Det 3 operators tried to process what those instructions meant for their immediate future. Not for the first time today they could not believe they were hearing the commands they were being issued. But the men complied with their orders and drug out the parachutes.

None of the Frogs were under the illusion this was a training exercise.

Col. Randal said, "Captain Jaxx was aboard a submarine that's been lost at sea. We have reason to believe he made it ashore and is currently somewhere at large on German-occupied Hydra. The Hudson will be arriving over the island in approximately forty-five minutes . . ."

The Frogs watched in disbelief as Col. Randal held up his right hand and touched the button on a deadly looking switchblade knife. *CLICK!*

"We're going to jump in, shoot everyone not wearing the same uniform we are, and recover him."

Then Col. Randal started cutting off the tape holding the splint to his leg.

8

SOMETHING STUPID IN PROGRESS

BEVERLY BLACKWELL HAD THE HUDSON DOWN LOW skimming the wave tops screaming through the late afternoon sky at War Emergency Power (WEP). Evening was fast approaching. Some said the scarlet and gold sunsets against the turquoise Aegean that turned to purple as the sun plopped into the sea were the most spectacular in the world. No one on the plane noticed.

In the sequence of events playing out across the AO the Hudson would arrive over Hydra shortly before nightfall. One hundred miles away Dr. Layton Winthrop was scheduled to land at the islet occupied by Long Range Desert Group Beach Watch Baker Team 6 shortly after nightfall aboard the MAS boat skippered by Brandy Seaborn. From there the decision had been made for his team to be transported to Sikinos by a U.S. Navy Landing Craft Rubber (Small) aka LCR(S) to link up with his agent. The rubber assault craft had recently been adopted by the LRDG due to its suitability for irregular operations. It was sturdy, portable (it could be transported, deflated, and then blown up with CO_2 bottles, hand pumps or air lines onboard surface ships) and they lacked a radar profile.

Lieutenant General "Geronimo" Joe McKoy would be standing off Sikinos in the High-Speed Launch with his team of OSS OG Det 3 operators awaiting developments.

Tonight's operation, while small in numbers, was arguably the most complex small unit action Raiding Forces had ever undertaken. Not only did it have a lot of moving parts, it required split-second timing and precision execution. A prisoner rescue or the extraction of an individual behind enemy lines has the lowest percent probability of success of any military action. To accomplish the mission it is necessary to have accurate intelligence on where the "precious cargo" will be at a certain time and a certain place. And that was just for starters. Detailed intel on enemy forces, weapons and equipment was a must. Once in possession of that information, then—and only then, can the mission commander develop a plan of action and conduct rehearsals. Extensive rehearsals are the secret sauce of small-scale operations and absolutely vital for rescues/extractions. Last of all it was a flunk if the precious cargo was killed in the process.

Dr. Winthrop did not have any of the required answers and there was no opportunity to conduct a rehearsal even if he did.

Onboard the Hudson Colonel John Randal was more in the dark than the professor. He had no idea where Captain Billy Jack Jaxx could be found except that he was believed to be on Hydra somewhere. He did not know the enemy strength on the island, what unit he would be facing, what weapons or equipment the Germans had or if any special equipment like heavy machine guns, mortars, etc., was present. And the jump the Frogs were preparing to make was unsanctioned, unauthorized and unofficial—not even known to be taking place by the staff at ABCHQ.

There was the distinct possibility OSS OG Det 3 would be dropping into an ambush. Had Vice Admiral Sir Randolph "Razor" Ransom learned of the plans for the Hydra jump prior to, he would have, in all probability, canceled it—one of Col. Randal's reasons for the secrecy and play-acting. He knew the Razor considered him too valuable a Special Forces commander to be lost on an ill-conceived, high-risk, forlorn hope.

Once the Frogs were chuted up and inspected by Lieutenants Chase Starrett and Ted Hamilton, Col. Randal stood in the aisle near the tail of the Hudson where everyone could see him. "You're going to drop on the main street of Hydra Port. This will be a low-level jump. . . five-hundred feet. Stay in a tight tuck position from the moment you exit the aircraft—out the door and down.

"There will be three teams, the Command Party, Lieutenant Starrett's team and Ensign Slade's team. In the event of being engaged on the DZ, utilize individual initiative—press the attack, be aggressive.

"Once Captain Jaxx has been recovered, Beverly will land. We'll reboard and the party's on when we get home. Officer's call as soon as the two teams are designated.

"Lieutenant Hamilton—it's time."

The Frogs watched riveted as The Great Teddy produced the flexible support brace Dr. Milam had provided in the event Col. Randal felt the urge to do "something stupid." It went on over his raiding boot and was adjusted to protect his knee. Then Lt. Hamilton produced a captured Fallschirmjäger's ribbed knee pad and pulled it on over the elastic support.

"Something stupid" appeared to be in progress.

Next Lt. Hamilton assisted in buckling the Colt 38 Supers around the Colonel's waist then helped him into a parachute, passing the leg straps from behind to be snapped into the chute's Quick Release System calling out, "Right leg strap, left leg strap."

Even though they were rookie one-jump troopers with less than a full day of Jump School behind them, the men were not so green they failed to observe Col. Randal was not bothering to wear a reserve parachute like they were. There must be a reason and it could not be good for them.

They were not wrong about that the drop was too low to deploy a reserve.

Could the Colonel be crazy enough to jump with his bad leg? It was said he was always first in the door. At this point, the idea of having to make their second parachute drop of the day began to sink in. A paratrooper's second jump is always the worst because of knowing what to anticipate. Ignorance is bliss—not on your second jump.

OSS OG Det 3's second was going to be on an enemy-held island during daylight hours which would have given even the most veteran paratrooper pause. The red light over the door flashed on. Ten minutes out. The Frogs were such novices that they had not yet learned the meaning of the onboard light system prior to the jump on Castelrozzo—but they knew now.

Col. Randal stamped his boot, being careful not to overdo it, held out ten fingers and gave the command, "Ten minutes!"

It was on.

Corporal Tom Murphy came out of the cockpit and made his way to the tail of the plane. What that indicated to Col. Randal was Beverly no longer required his services to navigate. She could execute the rest of the flight plan on her own. Murph the Surf plugged in a headset to be able to communicate with her. His role now was to retrieve the static lines once all the jumpers exited the aircraft.

After a commo check Cpl. Murphy said, "When I informed Miss Blackwell you would be leading stick sir, she made reference to your uncertain parentage and made a threat on your life."

Col. Randal said, "More information than I need, Murphy."

"Sir!"

Col. Randal ordered, "Officers report."

Lieutenant Chase Starrett and Ensign Westly Slade were briefing their teams. They broke away and shuffled toward the tail. Lt. Hamilton was already there talking to Waldo Treywick. The former ivory poacher was not thrilled to find out he would be jumping without a reserve because the altitude was too low for him to have time to deploy it in the event his main failed. The two composed Col. Randal's Command Party.

Waldo said, "How come the Frogs are wearing reserves?"

Col. Randal said, "For confidence."

"In what?"

Cpl. Murphy said, "Sir, Miss Blackwell reports fighting in progress in the village."

A high risk jump had just gotten more hazardous.

Col. Randal passed the word to his officers when they shuffled up. "We'll be dropping on a hot DZ. Have your men be prepared to go into action immediately upon landing. That means jumping with a magazine inserted in their weapon but *no* putting a round in the chamber until they hit the ground.

"Don't worry when there's no time to assemble your teams on the DZ—there won't be. The second you're down go straight over to the attack with the people immediately around you. Tell your men I want a maximum effort. Everyone fights. The only way off Hydra is to kill every Nazi on the island. Clear?"

"Clear, sir."

Col. Randal said, "For the record, gentlemen, if you ever conduct a mission as ill-conceived, poorly organized or recklessly executed as this one I'll have you reduced to the ranks, transferred out of Raiding Forces and assigned to the infantry.

"Let's do this."

CAPTAIN BILLY JACK JAXX HAD COMPLETED WORKING HIS way to the upper end of town stashing weapons where he saw fit. A forty-five-foot Motolance barge the Germans had captured from the Italians and pressed into service for inter-island troop transport was pulling into the entrance of the bay. It was slowly cruising toward the pier. He knew the barge was capable of transporting thirty PAX. Jack Cool hoped this one had not come calling loaded with a full complement of Nazis.

No other option being available that he could think of, Capt. Jaxx ran back down the street to his first fighting position. He had a 9mm Steyr MP-34 stashed there with spare magazines. It was in addition to the two slung over his shoulders. The hide was located off the end of the dock. This was what was known in tactical terms as a "Near Ambush."

The shallow draft barge slowly made its way to the pier. The landing craft's most distinguishable feature was an ungainly ramp elevated at a 45-degree angle sticking up above the craft's squared-off bow. The ramp was lowered by a long-necked winch to allow passengers and supplies to load and/or disembark. The configuration of the bow and gangway prevented Capt. Jaxx from being able to see how many troops were being transported in the open-air troop compartment.

But the barge being a Motolance class craft was the first piece of good news he had received the moment the police chief, who had since disappeared, casually mentioned the imminent arrival of 1st Mountain Division's replacement detail. Motolance barges were unarmed. Most of the larger German barges carried various types of heavy field artillery as expedient naval guns as well as an array of light and heavy antiaircraft machine guns which would have been a problem.

It would be ideal if Capt. Jaxx could draw first blood by engaging the Germans closed up forward preparing to disembark carrying their luggage like tourists. A mass casualty situation in the initial contact of an ambush initiated by an unexpected enemy would throw the Germans into confusion and give him cover to make a retrograde movement to his first fallback position. And it would not do much for the 1st Mountain Division troop's morale. The Germans were looking forward to a month of R&R, drinking, swimming, having their way with the local women and not taking an active part in the war. The Gebirgsjägers would definitely not be expecting to be ambushed.

But the Frankenstein-looking gangplank was preventing a clear field of fire. Maybe when the ramp was lowered it would no longer obstruct the view. That was exactly what happened except it still partially blocked Capt. Jaxx's line of sight. Not to matter. The instant it was winched down, but before the crowd could disembark, he commenced fire.

Not only did Capt. Jaxx mow down those Germans closed up waiting to step on the ramp aiming at the only visible part of them he could see—the head and shoulders. His 9mm rounds ricocheted off the bulkheads, buzzing around inside the close quarters of the troop compartment like a swarm of steel-jacketed yellow jackets and caused additional casualties. This was the classic textbook example of a chokepoint that channelized an enemy unit forcing it into the "Killing Zone" of an ambush.

The Nazis never saw it coming.

He was off to a good start, surprise being the most important element of a properly executed ambush, followed closely by superior firepower. However, given the odds it was going to be difficult for him to maintain fire superiority for long. If Capt. Jaxx had been a little closer he could have tossed in a couple of the potato masher grenades he had acquired earlier from 1st Mountain Division troops who did not need them anymore. That would have been perfect.

The firing, all his, seemed to go on forever, but in fact, he emptied three MP-34s, reloaded one and ran through its magazine again in under a minute—closer to thirty seconds. Capt. Jaxx was not the best of target shooters. Nor was he an accomplished long-range sniper. But on close in, moving targets, snap shooting with rifle, pistol or submachine gun he was world-class. Especially when engaging dangerous game like Nazis.

Capt. Jaxx would have stood his ground and kept fighting except the 1st Mountain Division mountain troopers toward the back of the troop compartment were starting to bail out over the sides of the barge into the water. How many was hard to determine. He knew those hardened combat veterans would make their way ashore and maneuver on him. That was the one thing he could not allow to happen.

Time to go. Capt. Jaxx grabbed the canvas pouches containing the spare 20-round MP-34 magazines and initiated what the tactics instructors at his old alma mater—the Infantry School at Fort Benning, Georgia aka "The Benning School for Boys"—referred to as a "tactical retrograde movement," He tended to think of it more in terms of hauling ass.

Capt. Jaxx made it to his next fallback position unscathed. The barge was still in easy SMG range. So he blazed away again hoping the Germans would believe there was more than one shooter in the village. That was not likely to happen. At least not for long considering the lengthy combat record of the 1st Mountain Division.

The Nazis began engaging, but it was of the "spray and pray" variety designed to suppress his fire. The Jägers only had a vague idea of his general location. They were hoping to cause him to shift his position again so they might be able to pick up on his location. This was a high-stakes cat-and-mouse game, Capt. Jaxx being the mouse—with no second-place winner.

The 1st Mountain Division veterans knew from long experience the best way to counter an ambush is to build up a base of return fire, achieve fire superiority and assault into it. That sounds simple enough in a classroom environment, but in actual practice, when caught in a well-prepared ambush with the ground littered with your dead and dying, almost nothing works. As one of Capt. Jaxx's OCS small unit tactics instructors had privately advised him on a smoke break, "Don't get caught in an ambush."

The 1st Mountain Division's relief detail had been ambushed—taken by complete surprise. The squad took heavy casualties in the opening seconds when it was initiated. Now the Germans were fighting back. Their only option was to battle their way out. The Gebirgsjägers needed to discover where their enemy was in order to concentrate their fires and bring their greater numbers to bear. Capt. Jaxx was not going to make it easy for them. By constantly shifting positions he was intending to frustrate the Germans' ability to counterattack.

The Jägers were fighting a will-o'-the-wisp who shot straight.

Capt. Jaxx was fully aware that if the unit being ambushed dramatically outnumbers the ambushers, as in this case, it can turn the tables. One small mistake on his part and it was curtains—no do-overs. He was headed toward his fourth fallback position, having darted across the street to the location where he had stashed one of the 7.62 SVT-38 semiautomatic rifles on the roof of a two-story building. From this new position to call his shots now he was going to have to partially expose himself when he fired. It had to be done because the distance of the contact was bordering on the effective range of the 9mm MP-34 SMGs.

Capt. Jaxx had just reached his next fighting position when the Hudson flew over.

COLONEL JOHN RANDAL SHOUTED AT THE STICK OF FROGS who were standing up and hooked up. "This is a short DZ. Stack the door hard. Come out tight and fast to reduce dispersion on the drop zone.

"CLOSE ON THE DOOR!"

Then he slapped his hands outside against the skin of the aircraft, knees bent, ready to launch. In this position he could not see the jump lights but he did not need to. Col. Randal was going the instant the Hudson approached dry land—exiting over water to land at the water's edge to allow for the aircraft's forward momentum and counting on a breeze blowing in from the sea to drift the parachutes. He had to time it right. Too soon he and all or part of the stick would land in the drink. Too late and they would miss the village entirely. A mistake in his timing and this mission would end right here.

Up ahead Col. Randal could see the tracers Beverly reported flashing back and forth. All green. His initial impression was the Germans were shooting at each other—Axis Forces used white and green tracers. Then the thought occurred that maybe Captain Billy Jack Jaxx had managed to capture Nazi weapons and was in a firefight. Knowing Jack Cool that was a distinct possibility.

Col. Randal could see the Motolance class barge docked at the pier. What did that mean? He had no idea. Beverly had come off War Emergency Power and cut back to 110 mph. If she was flying at too high a rate of speed when the stick exited her aircraft, the opening shock could and probably would cause the silk canopies to tear. The last thing any paratrooper wanted when jumping too low to deploy a reserve was a rip in his parachute.

Even with the reduction in airspeed, Hydra Port was coming up fast. If that was Capt. Jaxx fighting then Col. Randal had made the right decision to drop while there was still some daylight remaining. The original idea for a daylight drop was so the Captain could see them even if he was in hiding or from a distance. He would then be able to move to the DZ to effect link-up. And if there were Nazis in the village, as was now confirmed to be the case, the Frogs could close with them fast before the Germans had time to recover from the initial surprise of paratroopers making a forced entry assault landing unannounced and unexpected in their midst.

The Wehrmacht specialized in the art of counterattack. Possibly the best at it of any army in the world. Only a fool would give the Germans an opportunity to organize and launch one if he could help it. Jumping on top of them negated that . . . somewhat. The downside was parachuting into the center of a built-up area with unskilled paratroopers was a desperate move bordering on suicidal—even without a firefight in progress on the drop zone.

Hot DZs are never good.

Down below Col. Randal saw the water's edge rapidly approaching. Tracers were flashing back and forth in the village but the firing was uncoordinated—neither heavy nor disciplined. The situation on the ground was confused. He had no idea what was going on. This was not the typical meticulous operation Raiding Forces specialized in.

Arriving unannounced and unexpected was the only tactical principle of a small-scale operation he was complying with. No one in Hydra, Greeks or Germans, was going to be expecting Allied paratroops to drop from the sky. If that was Capt. Jaxx down there he would not be expecting them to come storming in "*Death From Above*" either.

Col. Randal knew surprise would only carry an operation like this one so far. Catchy regimental mottos wouldn't do it either. They were in for a battle.

The Frogs were not trained to fight battles . . . they were trained to blow stuff up.

Col. Randal leaned back inside to double-check to see if the stick had complied with the order, "Close on the door."

Totally unnecessary. Det 3 was stacked so tight if they got any closer they were going to shove him out of the plane. Lt. Starrett was struggling to hold them back. Prior to the commencement of the jump commands, Col. Randal asked Ensign Westly Slade to name the toughest Frog onboard.

Ens. Slade said, "That would be Lance Corporal Karlsson, sir—a Marine Raider. Used to play football for the University of Miami when he wasn't chasing women on South Beach or getting in bar fights. We call him 'Tank.'"

Col. Randal said, "Have him report to me."

Lance Corporal Ray "Tank" Karlsson shuffled up, "You wanted to see me, sir?"

Col. Randal said, "Ensign Slade tells me you're the Det 3 hard case, Corporal. I need you to be last jumper in the stick. You're going to be my pusher—Raiding Forces Rule Number 4: 'Right Man, Right Job.'"

"Pusher?"

Col. Randal said, "When I hit the silk, lower your shoulder and charge the door. We have to get out fast. Don't let anyone freeze or even slow down. Is that clear?"

"Can do, skipper." LCpl. Karlsson's crooked smile was the first sign of pleasure Col. Randal had observed on anyone since he ordered the Frogmen to reach under their seats and secure a parachute. It was not pretty.

The waterline was just off the toe of his right canvas-topped raiding boot. Beverly had the Hudson lined up on the street which in reality was not much more than a doublewide cobblestone sidewalk. She may not have liked dropping the OSS OG Det.3 Operational Swimmers because of their lack of experience and knowing the perils associated with jumping on a built up area but that did not keep her from following Col. Randal's orders to the letter.

Unable to shake the feeling he was making a big mistake, Col. Randal shouted, "Let's go!"

FIVE MILES OFF SIKINOS THE HIGH SPEED LAUNCH HOVE TO after a seventy-mile run at top speed. The skipper, a young Royal Navy Volunteer Reserve officer, Lieutenant Jeffery Featherstonehaugh, did not want to approach the island any closer until cloaked by the cover of darkness. Lieutenant General "Geronimo" Joe McKoy was of the same mind. He only had six Frogs to land ashore plus himself. And the professor's agent on the island had reported seventeen Criminals of the 999th Light Division stationed in the village.

Sparks, the radio operator—in the Royal Navy all radio operators were called "Sparks," came up to report he had made contact with Long Range Desert Patrol Beach Watch Baker Team 6. He was in receipt of a SHACKEL CODE message from Dr. Layton Winthrop. Decoded it read, "Intend to land with Command Party at 1930hrs. to link up with agent, locate pilot, secure same and come out via the LRDG TM 6 site to extract via Walrus. Stand by for follow-on message once we access situation."

This was a longer transmission than desirable but since radio was the only means to coordinate a fast-developing operation with little advance intelligence, it was necessary so all parties knew what the other was doing.

Lt. Gen. McKoy said, "Respond with a Wilco."

Apparently, the professor had modified the plan. Instead of loading the pilot, his team and the LRDG team on Brandy's MAS boat for the return run to Levant Schooner 7, he had decided to have Wing Commander Paddy Wilcox fly them directly from the Beach Watch position to ABC. Lt. Gen. McKoy concurred with the change. It got the doctor's party and the LRDG away from the vicinity of Sikinos and back to base faster.

The six OSS OG Det 3 operators gathered around Lt. Gen. McKoy, eager to be kept in the loop. The Frogs were too new to Raiding Forces to realize the troops were never left in the dark once an operation commenced, though not all the details of every mission were briefed until the last minute.

Lt. Gen. McKoy gave them an update on the change of plans. Then he said, "In addition to the word from Dr. Winthrop I have a flash message from Beverly. Colonel Randal, Lieutenant Hamilton, Mr. Treywick and the rest of Det 3 parachuted onto Hydra to rescue Captain Jaxx who was shipwrecked there when the submarine he was on got itself sunk. They're fighting at this time."

This was a stunning announcement.

Lt. Gen. McKoy said, "Catch a few Zs, boys. Nothin's happenin' here for a couple a' hours. I'll tell you what's next when I know. The situation'll be changin' minute by minute. Don't be expectin' any five-paragraph field orders from here on in. Stay loose—we're wingin' it."

The Frogs laughed but the sound had a nervous ring to it.

Lt. Gen. McKoy said, "I'm a' the opinion we're in for a serious firefight before this night is over. Make sure those Marlin guns a' yours are oiled and ready. I want to see what they've got when you light 'em up."

The Det 3 operators went below to rack out. The former Arizona Ranger, Mexican Incursion Chief-of-Scouts, U.S. Marshal and Medal of Honor recipient made the operation sound so . . . ordinary. The OSS Operational Swimmers were more than a little curious about how many small unit actions and big battles Lt. Gen. McKoy had been in.

What the Frogs *really* wanted to know was how many men he had killed.

WONDERING WHAT MAJOR THE LADY JANE SEABORN might be doing right about now, Colonel John Randal made a vigorous exit in the classic tight tuck body position, but not having a reserve strapped to his chest to hold between his hands the jump felt awkward. Muscle memory is for times like this—under extreme duress. That is why repetition, repetition, repetition in training is so important. With his elbows pulled in tight against his sides, he found himself momentarily clasping for a nonexistent reserve, which meant he was well-trained. He would have thought it funny had it not been for the German soldiers down below in and around the barge gawking up at him—that removed the element of humor from the equation.

Unknown to Col. Randal, a serious number of casualties had already been inflicted on the troops disembarking the Motolance. If it were not for that and the sudden shock of seeing paratroopers storming out of the plane at low altitude, the Nazis would have had a turkey shoot of the type he and Captain Billy Jack Jaxx did on Leros. But too many things were coming at them all at once for the overwhelmed Germans to react in an organized military manner.

Onboard the Hudson, when the command “Let’s Go” rang out, Lance Corporal Ray Karlsson aka “Tank” lowered his shoulder, put it against the parachute pack containing the X-chute of the man in front of him, and charged the door. A former right guard for the Miami Hurricanes, he was tailor-made for the role of pusher. In the rush to get out of the plane the stick abandoned the Airborne Shuffle. Frogs were literally running out the door. No tight tuck body position—just charging out into space with their legs bicycling. The jumper in front of LCpl. Karlsson stumbled, lost his balance and went down. There was a reason paratroopers onboard an aircraft-in-flight were taught the Airborne Shuffle—never pick up your feet to avoid stumbling. Without missing a beat, Tank reached down, grabbed his teammate by the collar with one hand, dragged him to the exit and jumped, taking the Frog with him. As far as clearing the airplane went, no one could have done it any better—except for the body positions on the exits. Those were awful.

Capt. Jaxx was on the roof of a two-story building sniping at muzzle flashes with the big 7.62 Russian SVT-38 rifle when he saw the Hudson line up on the village street and commence its run. It was flying straight at him due to the slope. The airplane was so slow, low, and so close he could recognize Beverly Blackwell in the cockpit. She was coming on strong.

What was this?

Then jumpers began exiting. Capt. Jaxx was so startled he briefly checked fire at the exact time he should have been putting out suppressive fire to cover the drop. Fortunately, the Germans were taken off guard as well and momentarily ceased fighting to try to get a grasp on what might be taking place. Their world was turned upside down by having to fight their way ashore on what for them was supposed to amount to a thirty-day paid vacation. Then enemy paratroops started dropping with evil intent. Even for veterans as experienced as the 1st Mountain Division troops, the situation was an overload to process.

Col. Randal came down right in front of the barge exactly where he intended. There were dead and wounded Nazis scattered on the ground and in the water. Germans who had been unable to disembark and were unwilling to go over the side were still onboard. They were returning fire uphill toward the far end of the village. Those men who had elected to abandon ship and make for the shore were taking whatever cover could be found along the shoreline and firing also. Two had drowned, weighted down by their equipment. When any of the Germans raised up to engage the paratroopers storming down, a shot rang out from up the street and dropped them.

Not able to see all that was taking place because he was blazing in backward, Col. Randal came in hot with his head down watching the ground whizz past out of his peripheral vision. He swiveled the instant his boots touched the ground and hit all five of his points of contact in rapid succession. Then he hammered the QRS, dropping his parachute before he quit rolling. It was one of his better PLFs and almost pain-free. The operative word being *almost*.

Automatically with no conscious thought, acting out of pure reflex, Col. Randal reached for a U.S. Army issue Mk II Fragmentation Grenade even before he recovered—pulled the pin and threw it into the Motolance. Fuel or ammunition, possibly both, must have been stored aboard the barge. The grenade went off followed by a sharp secondary explosion complete with leaping flames and a perfect white smoke ring shooting up into the sky. That was surreal.

Frogs were coming down, landing in the street, on rooftops, in alleyways, smashing into walls. The DZ stretched the length of the street. It was a perfect drop on a nightmare DZ followed by some of the worst PLFs ever performed. It was probably a good thing the OSS OG Det 3 operators did not know how to evaluate drop zones. Landing on a parade ground would have seemed like a bad dream to them.

Strung out as they were, thin on the ground everywhere, the jumpers were vulnerable. The Germans instantly recognized an opportunity. The 1st Mountain Division troops who had made it off the barge were not long in getting a counterattack going. Looping around behind the buildings lining the street on the right flank they were in position to screen their approach as they attempted to envelop the drop zone. Meanwhile, every Nazi not part of the flanking element was now laying down a base of fire to either pin down the paratroopers on the DZ or kill them. This is called "Fire and Maneuver"—a basic infantry tactic known in all armies throughout history—and is very effective. Col. Randal had wanted *his* people to go over to Fire and Maneuver immediately upon landing but the Germans beat them to it.

The momentum of the airborne assault was beginning to spend itself.

The 1st Mountain Division Jägers were not only maneuvering on the DZ they were achieving fire superiority. This was aided by the fact that while arguably the most highly-trained specialist troops in the U.S. Military system, the OSS OG Det 3 Operational Swimmers were lacking in basic infantry skills. The situation was not helped by their having made a previous parachute drop earlier in the day that resulted in almost every Frog incurring some injury. The same happened on this one. Virtually all the OSS jumpers were banged up for a second time.

Playing hurt, they were slow going into action.

Two Nazis—seeing Col. Randal was not carrying a rifle or submachine gun, likely believing him to be unarmed—came out of nowhere and charged him with bayonets affixed to their Mauser 98K rifles. Few things are more unpleasant than to be assaulted by enemy soldiers at point-blank range with bayoneted rifles in the "On Guard" position, knowing they were determined to run you through with ten inches of cold steel.

Col. Randal dropped them both rapid fire with a pair of rounds each center of mass with one of his Colt .38 Supers. The Germans should have never broken cover. They should have just shot him. Even battle-hardened veterans of a half-dozen major campaigns can become rattled and make a mistake in fast-developing, close-quarters combat like what was taking place in Hydra.

Unknown to Col. Randal there were fourteen Germans on the barge when it arrived at Hydra. He jumped in with a twelve-man stick. The sniper, who might or might not be Jack Cool, had taken out nearly a third of the Nazis before the drop. And the grenade he threw into the barge had killed or incapacitated an unknown number more. So the odds were swinging in favor of OSS OG Det 3 and company even though the momentum had shifted to the 1st Mountain Division.

The problem for Col. Randal was numbers do not trump firepower or infantry fighting skills. With some of the Frogs still not out of their chutes and others dazed due to their high-impact landings incidental to the low-level jump, the German maneuver element was about be in a position to overrun the DZ. And the covering fire from the Nazi support element was increasing in volume to the point the Jägers were on the verge of attaining fire superiority.

To their credit, more and more of the Frogs were beginning to engage. However, the firefight was spiraling out of control. Col. Randal knew it would be all over if the Germans were able to complete their envelopment and assault through the drop zone. Powerless to prevent that in the few seconds he had to work with, Col. Randal realized his people were in serious trouble.

Then the Germans made a mistake. The maneuver element moving up behind the buildings got in a hurry. They failed to maintain proper five-yard spacing (at minimum) and became bunched up in the rush to close with the parachutists before they could get assembled. What the Nazis were intent on was called "Violence of Action" and it wins battles. However, it is never a good idea to violate the basic principle of dispersion on the battlefield, meaning maintaining a safe distance between men, when making a movement to contact.

The Hudson roared over at an extremely low level, less than tree top level, had there been any trees. Beverly was reinterpreting her orders to racetrack off Hydra Port—considering them as more of a suggestion than a direct order. She had four organic .50 caliber AN/M2 Browning machine guns on her aircraft. They were all controlled by a firing button on the steering yoke. Two were mounted in the nose and two were in the Paul Boulton turret on top of the fuselage. The Browning HMGs were upgrades from the .303 caliber Vickers K machine guns the plane was armed with when it arrived from the factory. The Hudson's firepower was equal to or superior to most RAF fighters. All guns were calibrated to converge straight ahead two hundred feet out. Different flyers adjusted their machine guns to converge at different ranges from 200 feet to 1200 feet depending on personal preference—Beverly liked close. Each AN/M2 had a cyclic rate of between 750–850 rpm. Times four that was a lot of monster-sized armor piercing, incendiary and tracer HMG rounds tearing up a target—56 per second, roughly.

Beverly caught the Germans running behind the buildings racing to flank the drop zone in the open but bunched up in their eagerness to close with the enemy. The four Browning's made a sort of yawning sound as .50-caliber rounds the size of cigars walked through the clump of enemy troops. *YAAAAAMPH.*

And it was over. Just like that. In seconds.

Seeing what occurred, the Germans in the port area started waving white handkerchiefs and calling out "*Nicht Schiessen*!" The 1st Mountain Division veterans knew when to call it quits. They did not want any part of being strafed by .50 calibers.

The Hudson was gone, banking hard right to come back around for another gun run. Beverly might have been a former University of Texas beauty queen, a member of Tri-Delta Sorority and because of her accent considered by some to be a dizzy blonde. But those people had never served with her. If there was a more valuable member of Raiding Forces, Col. Randal did not know who it might be. She definitely was this day.

Bleeding from a gash on his forehead, Ensign Westly Slade jogged up to report, "Sir, I have two KIA and the rest of my team is pretty dinged up." He had just lost his first troops in battle. Never a good experience for a small unit commander. One he would never forget.

Col. Randal said, "How?"

Ens. Slade reported, "Thibodeau, our Cajun, was shot in the air coming down. Smith broke his neck on landing, sir."

Lieutenant Chase Starrett arrived as composed as if he were on a training exercise. "I have one man with a broken arm—compound fracture. The rest of my team is barely ambulatory, sir."

The Hudson came back over for a second gun run and the big fifties went to work again. This served to increase the intensity of white flag waving from the Motolance. The Germans remaining on the barge did not want Beverly strafing the dock area though how she could accomplish that without hitting OSS OG Det 3 was open to question. The Nazis were not taking any chances.

Col. Randal said, “Ensign Slade, take your team, work your way up the street and mop up any Germans remaining. Probably a few hiding in the buildings you’ll have to clear. Don’t take unnecessary chances. Now would be the time to put your demolitions skills to use. Anyone engages, blow the building down—let’s see what you’ve got.”

“Lieutenant Starrett, secure the pier and the barge. Assemble all prisoners on the Motolance. Any give you trouble—light ’em up on the spot. Stand ready to send half your team to Ensign Slade’s assistance on request.”

Both officers responded, “Yes, sir!” Then trotted off to pick up their troops and move out on their individual assigned missions.

Col. Randal already knew Lt. Starrett was a stud. Ens. Slade appeared to have the makings of one. He was more than satisfied with the performance of both his young officers and the men of OSS OG Det 3. Lt. Starrett for leading troops he barely knew and who did not know him well. Ensign Slade for performing as difficult a task as the military could throw at him without being adequately trained for it. As for the Frogs, they had done everything asked of them without complaint.

Col. Randal spoke to LCpl. Karlsson as he trudged past looking like he had landed face first and skidded on the cobblestone street, “Nice job, Tank. You can be the pusher on my stick anytime.”

“Thanks, skipper . . . OK if I complete parachute training first?”

“We’ll see what we can do about that.”

Then Col. Randal led his Command Party—consisting of Waldo Treywick and Lieutenant Ted Hamilton—up the street traveling troops shaken out in a loose half-diamond formation. They moved in the direction from where the sniper fire directed at the Germans on the barge had been originating. Capt. Jaxx met them halfway armed to the teeth with captured German weapons. He did not seem much the worse for wear and tear for his adventures.

Capt. Jaxx said, “Got to be the gutsiest jump in Airborne history, sir.”

Col. Randal said, “Enjoy your time in the Submarine Service, Jack?”

BRANDY SEABORN CUT THE MOTOR ON THE MAS BOAT TO dead slow for the final approach to the uninhabited islet occupied by Long Range Desert Patrol Beach Watch Baker Team 6. The exfiltration plan had changed. Now instead of standing by with the LRDG, she was going to drop off Doctor Layton Winthrop and his team and return to LSF 9 to await developments. Instead of the MAS boat transporting Dr. Winthrop's party and the LRDG team back to ABC, Wing Commander Paddy Wilcox would fly them back aboard the Walrus.

No plan is carved in stone. At least not in Raiding Forces. Deviations being announced by frag order which is what they are designed for—changes of mission. With an operation in progress adjustments to the original plan were something to be expected. Colonel John Randal required his officers, NCOs and the troops they commanded to adapt to changes to the mission profile in the field and continue to march without any interruption in the operational rhythm.

It is often said, "No plan survives the first contact with the enemy." That is not true. A change-of-mission is common, up to and including the initiation of the execution phase of any operation. The ability to make rapid adjustments to the scheme of maneuver while on the move and continue the mission was a must in Raiding Forces. Not every combat arms professional, officer, NCO or EM has the personality, mindset or mental agility to be able to adapt to sudden deviations in the plan without becoming frustrated—that last being critical.

Raiding Forces called it "rolling with the flow."

Not having had the benefit of much time for the all-important "prior planning" before his team launched from ABC, Dr. Winthrop had been putting the flight time to the target to good use, running through possible combinations of options available to him. While he could only come up with minor tweaks the changes were designed to provide the mission its best possible chance for a successful outcome.

The LRDG patrolmen signaled the MAS boat with a red-filtered flashlight. Brandy idled to the beach with barely a bump. Dr. Winthrop's party disembarked immediately. The professor had decided he would leave Beach Watch Team 6 behind on the spit. He could move faster with just his party. He wanted a quick in and out.

Dr. Winthrop's revised plan was simple. As soon as Private Norvel Hansen completed a radio check with Levant Schooner 7 and the High Speed Launch standing off Sikinos, the Life Boat Service Men (LBSM) would paddle the professor's team to the island. The two LBSM would set up an Extraction Rally Point and remain on the beach to secure the LCR(S) while the doctor, Lovat Scouts and PFC Hansen went inland to link up with the SOE agent at his home.

The premise of the rescue centered around the possibility the Sikinos contact was sheltering the pilot, Flying Officer Bash Potgieter, or knowing where he was located. If that turned out not to be the case, then Dr. Winthrop would have to go into the village and find someone who did.

The one advantage this mission had going for it was the occupying troops on Sikinos were from the 999th Light Division. There was a certain sameness about the way the Germans carried out their occupation duties on the small Aegean islands, particularly the 999th. Typically at night the Criminals would all be concentrated in a local bar they had commandeered for their private use for safety's sake with some bold souls in private residences scattered throughout the village sleeping with a local woman, her daughter or possibly both. The Nazis would not be out and about now that it was dark. They were afraid of the local Greek men who would kill them if they could.

That predictability accorded Dr. Winthrop freedom of movement.

PFC Hansen reported, "Sir, I am in contact with General McKoy. He's standing by awaiting your instructions."

Col. Randal had ordered Horn Dog to render the professor the same degree of military courtesy as he would Vice Admiral Sir Randolph "Razor" Ransom or Lieutenant General "Geronimo" Joe McKoy. PFC Hansen was complying. He could soldier when he wanted to.

Dr. Winthrop said, "Advise him to close to just offshore and continue to stand by."

"Roger that, sir."

The two-mile paddle to Sikinos did not take long. Life Boat Service Men are arguably the finest rough water small boat handlers in the world. Tonight the Aegean was as smooth as silk. They made good time.

Paddling ashore in the dark of night aboard a twelve-foot long, five-foot-eleven-inch wide rubber raft on a German-occupied island was always a high drama event. Even when there was very little likelihood of enemy contact.

You never knew.

Dr. Winthrop spoke to the LBSM briefly to coordinate. Then he whispered to the Lovat Scouts, "Move out lads. I shall keep you on azimuth."

Scouts Munro Ferguson and Lionel Fenwick stepped off at a pace the military likes to describe as "smartly." Dr. Winthrop had heard about the Scouts' ability to operate at night. Even so, he was impressed with the ease and speed of their movement over unfamiliar terrain in the dark. They were like cats.

The Doctor's SOE contact lived on the edge of town a little over a mile from where the LCR(S) had landed. He should be at home. Greeks had a curfew that started at nightfall. Anyone caught violating it was shot precipitously—*BANG*, no questions asked.

The patrol closed the distance to the agent's home in short order and went into a tight perimeter. Dr. Winthrop advanced alone to make contact. This was as simple as knocking on the door and giving a preestablished authenticator.

The Greek, known only as Artemus, was cautious, as he well might be, but he opened the door. The professor had studied Raiding Forces Rules and now he stuck with Number 2 which seemed to work in most instances—"keep it short and simple."

"Where do we find the pilot who was shot down?"

"The Germans are holding him at the police station."

This was not the answer Dr. Winthrop was hoping for, fully aware, as Col. Randal was famously known for saying, "Hope is not a course of action."

"I need you to take us there."

Artemus did not seem overjoyed by the prospect. He was paid by SOE to report information on enemy activity on Sikinos—not to be a man of action. After the professor and his men departed the island he was going to be left behind. There would be consequences.

Dr. Winthrop turned and said quietly, "Horn Dog."

PFC Hansen trotted up to the house.

Dr. Winthrop got on the radio and explained the situation to Lt. Gen. McKoy who replied, "Give me an up after you've secured the precious cargo. When you do, we'll go ashore and put in a diversion to cover your exfil, over."

"Roger."

Doctor Winthrop was the perfect model of a British academician at war . . . scholar, soldier, spy. He calmly shifted gears when the parameters of the mission changed from being a simple recovery to a hard target rescue. Returning to ABC without the pilot he came for was not an option the professor even considered.

Artemus led the way while the Scouts shadowed his every move. They had privately recommended to Dr. Winthrop that a rope be attached to the agent. If he slipped away for whatever reason their mission was going to become considerably more complicated or possibly even compromised. The professor declined the suggestion though admitting to himself it was an eminently sensible precaution.

Artemus lived at the top of the escarpment above Chora, the capital, that sloped all the way down to the bay. He led the team down through the village keeping to the main street. The population of the island had been skewed when the Greek government deported a thousand or so communists to Sikinos prior to the war. Now the newcomers outnumbered the locals nearly four to one.

Most of the deportees did not live in the village, being forced to eke out a meager existence by farming in the interior on land not ideal for growing crops. The exiles were considered outcasts by both the Greek islanders and the Germans. The Greeks hated the Communists for overpopulating their island and the 999th had no more compulsion about killing them than they did vermin—their politics making them the natural enemy of the Third Reich. No one in the Axis chair-of-command would question the Criminals' actions.

It only mattered to Dr. Winthrop because the politics on the island made it unlikely the deportees would interfere with his patrol.

Dr. Winthrop's team patrolled down the main street. Their rubber-soled raiding boots were virtually silent on the cobblestones. If anyone in the houses heard them as they ghosted past, they acted as if they did not. It paid to mind your own business in Chora. Whoever was out there could be a German patrol, a band of local Greek men hoping to encounter a drunk Nazi foolish enough to venture out alone or a gang of roving Communists attempting to steal food. After dark Sikinos became a sinister place.

Everyone in the party—with the exception of Artemus—was armed with a .30 M1 Carbine. The patrol was on full alert with their weapons at the high port as they traveled. Police stations were one of the three municipal buildings found on most islands. They were usually located on the main street. Bars were generally on side streets out of some odd sense of decorum. On islands all across the Aegean the locals drank like fish but they chose not to advertise it. The Greek Orthodox Church, while not opposed to the consumption of alcohol, frowned on immoderate indulging of spirits. What that meant for the operation was the German occupation troops would not be in the immediate vicinity when Dr. Winthrop raided the police station.

While none of the residences or commercial buildings they slipped past had lights turned on, most of the houses had candle lamps burning. Patrolling through an enemy-occupied village at night was always a surreal experience. The mellow candlelight made you wonder who was watching. No way to know. But there was no way of shaking the feeling someone was.

Artemus came to a halt and pointed. Directly to the patrol's left front was a small stone building. A mellow candlelight was showing from inside. Dr. Winthrop sent the Lovat Scouts forward to recon. One second they were there and the next they were gone as if The Great Teddy had waved his magic wand and said, "Hey, Presto!" The professor was grateful the Colonel had seen fit to include the two Scouts on his team. Normally the Lovats worked for the Colonel exclusively.

Time seemed to drag. That was only an illusion. The Scouts were back in under three minutes—poof, they simply appeared.

Scout Ferguson whispered, “A couple of Jerries inside drinking and playing cards. The pilot is in the cell in the back of the room. Three windows—one in front, one on each side of the building—all open. No back door.”

Dr. Winthrop said, “How do you recommend we proceed?”

Scout Fenwick said, “Attempt to do this as quietly as possible, sir. I go in while Munroe covers me with his suppressed pistol from one of the side windows. Horn Dog provides additional security from the other side window. Since he does not have a silenced weapon he will only engage in an emergency. You and Artemus stand fast and cover the front window and the door.”

Dr. Winthrop noted this was the most he had ever heard either of the normally taciturn Lovat Scouts speak. He also recognized sound advice when he heard it. Fenwick clearly knew what he was doing.

“Brief Ferguson and Hansen, then move out.”

After a short conversation with PFC Hansen and Scout Ferguson the two disappeared around the sides of the police station. It was a small building, approximately 500 or 600 square feet, like police stations on most other minor islands. It did not take the two men long to get into position. Each signaled when they were in place by slapping the wooden butt of his M1 Carbine twice with the flat of his hand.

When both signals were heard, Scout Fenwick passed his M1 Carbine to Dr. Winthrop to secure while he made entry. Then with his suppressed .22 High Standard Military Model D at the ready, he walked up to the door, and as a mere formality tried the knob. Not locked—a surprise. He had been ready to kick it down.

The 999th held the Greek islanders in contempt for being Greeks and for allowing them to sleep with their women—as if they had any choice. Holding your enemy in contempt, no matter how weak or unwarlike they may be, is a recipe for disaster sooner or later. Other than not going out after dark, except in large groups, the Criminals never practiced even basic security measures. Why should they? They were on vacation. The civilian population was cowed.

Scout Fenwick stepped inside and with a pair of steps closed the distance to the card table to virtual contact range. The two drunk Nazis looked up unalarmed to see who it might be walking in.

He put a .22 round in each man's head. *WHIIIIIICH, WHIIIIIICH!*

Both Criminals had startled expressions on their faces as the tiny bullets slammed home. They were dead in an instant or at least comatose preparatory to being dead. One fell over the cards on the table and the other leaned sideways then slumped to the floor.

Col. Randal's standing order for the use of suppressed .22 was, "One to the head two to the body."

WHIIIIIICH, WHIIIIIICH, WHIIIIIICH, WHIIIIIICH!

Scout Fenwick called, "Clear!"

From his position covering the door the professor never heard a sound louder than the sea breeze prior to Scout Fenwick's signal.

A few minutes later Scout Fenwick called out again, "Two coming your way."

RAF Flying Officer Bash Potgieter was literally wild with joy to be rescued. Though it was somewhat muted because of his battered physical condition. It was his opinion, which he was trying his best to share with the Scouts . . . babbling away as they moved him outside, that the 999th Criminals needed to be locked in an institution for the feloniously insane. His guards had made it clear they were planning to tie him to a post in the morning, soak him in gasoline and light it off as entertainment for the rest of the detachment. As soon, that is, as they recovered from their hangovers. Like all combatants, the South African pilot had imagined any number of ways he could get killed in the war. Being burned at the stake was not one of them.

Dr. Winthrop ordered, "Recover" just loud enough for Scout Ferguson and PFC Hansen to hear. When they appeared from around the sides of the police station it was time to go.

PFC Hansen squeezed the push-to-talk switch on the radio's handset and said, "Execute, Execute, Execute."

Lt. Gen. McKoy responded, "Roger that, out."

9

AIN'T NO SUCH THING AS OVERKILL

LIEUTENANT GENERAL "GERONIMO" JOE MCKOY assembled his OSS OG Det 3 Operational Swimmers on the stern of the HSL. The boat was lying a quarter mile off Chora but was starting to close on the island. The men were wired tight. It had been a long day and now they were about to invade a place where there was almost zero intelligence available other than it was occupied by a reinforced squad-sized detachment of the 999th Light Division. No one seemed to know what that meant exactly.

Lt. Gen. McKoy said, "OK boys, listen up. Here's the deal. We're gettin' ready to create a diversion to cover Doctor Winthrop's withdrawal—he's got the precious cargo we came for. I'm goin' to let you vote. We can land ashore, wander around and shoot the place up for fifteen, twenty minutes. Or we can locate the bar the 999th Criminals'll be holed up in like they always do and you can blow it up."

The Frogs looked at each other incredulously. Was the General joking? Vote? On tactics? In the middle of a mission?

Lt. Gen. McKoy said, "All in favor a' shootin' up the town, raise your hand."

No one responded.

"All in favor a' blowing up a bar full a' drunk Nazi Germans, raise your hand."

Feeling foolish, the OSS OG Det 3 Operational Swimmers raised their hands, sort of—Frogs like to blow stuff up.

Lt. Gen. McKoy said, "That's too bad. . . bein' as how I'm more of a 'roll into town, shoot the lights out and then adios' in a cloud a' dust like the good ol' days kinda guy. So, here's what's goin' to happen. Sub-Lieutenant Grey speaks Greek. We'll take him along with us when we go ashore.

"At the first house we come to we'll grab whoever answers the door and the Lieutenant'll convince him to take us to the 999th's hangout. May have to have a little help with that part. The Greek might be a little reluctant to get involved us being strangers but we'll get 'er done. When we locate the Criminals' hangout you boys do your thing and blow the place up. Petty Officer Nelson, you're senior man. Make it happen. Now hear this . . . I want a big-time bang."

Petty Officer 3rd Class Joe Nelson said, "Can do, sir."

Lt. Gen. McKoy said, "Understand, the purpose of the exercise is to cover Dr. Winthrop's exfil. He's in the process a' makin' his way outta town as we speak. We do our part, divert the opposition long enough for his team to make a clean getaway, and it's mission accomplished.

"As a bonus we'll shoot up the town in an orderly fashion durin' the pullback to the boat for the benefit a' those Criminals sleepin' in local houses. Give 'em somethin' to think about and let me have a little fun. We're goin' ashore in zero five.

"What are your questions?"

Typically when issued an order if the Det 3 operators did not have a specific question that would advance their understanding of the mission they did not ask. The Frogs were disciplined about that. They had been well schooled on the proper procedure for receiving an order. Hold all questions to the end and do not ask unless there was a reason to clear up something.

PO3 Nelson assembled the team. While they may not have been trained as combat infantry they were arguably the world's best naval demolitions men. Each carried a small demo kit at all times. In them, among other things, was a pound of C-3 and a ten-foot coil of detonation cord. Spliced all together with the blocks of explosives the team's det cord would result in a sixty-foot-long ring main system. The Frogs would be able to string it around all or part of the bar with a pound of C-3 spaced every ten feet. It should be enough to do major structural damage to the building and the Criminals drinking inside. However, Lt. Gen. McKoy wanted a massive explosion. And *that* the demolitions in their demo kits would not deliver, even with the extra C-3 the Frogs had added prior to LD Time being adherents of the P-for-Plenty Formula.

PO3 Nelson said, "Ideas anybody?"

Which is exactly what a good leader does when he does not have a clue what to do in a given situation—ask for suggestions.

PFC David Schwartz said, "The HSL has a pair of roll-off Mark VI depth charges. One of those should do the trick."

The depth charges were a weapons add-on Vice Admiral Sir Randolph "Razor" Ransom had ordered for the Small Raids Inc. fleet of High-Speed Launches—an easy install. It was the first of several armament upgrades he had in mind for the HSLs. The Frogs were familiar with the various organic weapons they might find on small craft likely to transport them. They knew the specifications of the Mark VI.

PO3 Nelson said, "How are we supposed to move a 300-pound depth charge to our target? We don't even know where that is."

Private First Class Bray Landreth said, "No problem. We'll lash together a couple of the HSL's twelve-foot boathooks and drag that bad boy."

PO3 Nelson said, "Works for me. Do it. Let's go, shake it up, boys! The General doesn't strike me as the type who likes to be kept waiting around."

Lt. Gen. McKoy was observing the Frogs with one of Waldo's unlit cigars in his teeth, well satisfied. PO3 Nelson exhibited exemplary leadership problem-solving traits. Not afraid to ask for ideas from his men when he did not have the answer himself. And decisiveness. Geronimo Joe was impressed. Enough so that on return to ABC he would be recommending to Colonel John Randal the Petty Officer, an enlisted grade in the U.S. Navy equivalent to an Army corporal, be considered for the Raiding Forces Officer Candidate Program. Subject to, that is, how well he handled the Frog team once they landed ashore.

It took more to qualify as a Raiding Forces officer than being a good problem solver. At a minimum to meet standards an officer had to be able to craft a plan or evaluate one issued to him, organize his troops to conform to the mission, issue Warning and Operations Orders, maintain his command presence at all times—which is an art more than a science—lead by example while keeping his troops under tight control and *execute* the plan. Sometimes while on the move with very little intelligence to go on while being shot at . . . or at least with the possibility of being shot at.

It is harder than it sounds and there was no way to know in advance who had the right stuff. It was not always who you might imagine—good leaders not necessarily being movie-star handsome or over six feet tall with bulging muscles.

DR. LAYTON WINTHROP AND HIS TEAM WERE EXITING THE town. They had to escort Artemus back to his house before cutting cross-country to link up with the Life Boat Service Men and the Landing Craft Rubber (Small). That was the promise he made to his agent when he came knocking on his door. If he wanted to keep him in place and have him continue to work as his man on Sikinos he was going to have to honor it. The problem was that doing so substantially increased the distance the team would have to travel to reach the LBSM waiting to extract them.

Flying Officer Bash Potgieter was coughing up blood. Not a lot but enough to be concerned. He had been injured when his Spitfire was shot down and roughly handled by the 999th Criminals after his capture. The nearly three miles the team would have to cover to the Extraction Point did not sound like much but tonight it was. Due to the physical condition of their precious cargo and the rugged nature of the island's terrain the team was not going to be able to move as fast as they needed. They would likely be forced to take breaks from time to time to allow the pilot a chance to catch his breath.

In the event not all the Germans were drinking themselves into oblivion and somehow discovered the two dead Jägers at the Police Station and the prisoner missing, making it to the extraction point would become problematic. The Criminals would cease their partying and drunk or not come after them hard. In the German Army losing a high-value prisoner could, and probably would, result in the senior man being shot. Likely the entire detachment. There was no shortage of convicts in prison to replace them.

Dr. Winthrop said, “Artemus, can you locate a mule for FO Potgieter to ride?”

Artemus said, “I shall try, Doctor, but no one is going to be eager to have aided in this man’s escape. The Germans will view anyone who does as being an accessory. Punishable by death for him and his entire family. Possibly his neighbors as well.”

Dr. Winthrop said, “Make the attempt. I am not sure the Flying Officer can make it under his own power. If we have to carry him that shall dangerously slow us down. We need to complete our exfiltrate prior to sunrise.”

Artemus said, “The farms in the immediate district are all worked by exiled Communists. Attempts to cultivate a relationship with them have met with no great success. They hate all Greeks loyal to the government as well as Italians and Germans.”

Dr. Winthrop said, “How do they feel about South Africans?”

ON HYDRA COLONEL JOHN RANDAL, CAPTAIN BILLY JACK Jaxx, Waldo Treywick and Lieutenant Ted Hamilton aka “The Great Teddy” were backing up Ensign Westly Slade and the only two surviving Frogs on his team able to fight. They were preparing to clear the remaining 1st Mountain Division Gebirgsjägers from the building they had been skirting in their rush to envelop the DZ when Beverly

Blackwell strafed them to good effect. Exactly how many Germans were inside was an unknown but it could not be many. It was impossible to get a good count in the confusion of the firefight and the rapidly changing events. To complicate things, the number of Nazis on the barge when it arrived was undetermined, as was the size of the detachment they were relieving. A squad has a Table of Organization & Equipment (TO&E), a detachment was an ad hoc formation and could contain any number of troops. And in addition different types of divisions, infantry, mountain, airborne etc. had different sizes of divisional assets.

Col. Randal needed to mop up all of the remaining enemy troops. He did not want any of them to be able to come out of hiding and take the Hudson under fire as his people were loading for the flight back to ABC. One or two well-trained marksmen armed with automatic weapons could turn the "getting the hell out of Dodge" phase into a disaster. Missions have to be thought through all the way to the end—extractions are highly vulnerable to counterattack.

A hot exfil is worse than a hot forced entry assault.

Col. Randal was letting Ens. Slade retain command of clearing the buildings. His Command Party was acting as the support element. The idea was to pin down the Nazis inside while a demo team moved up and placed charges. The Frogs had a good amount of C-3, having redistributed the explosives belonging to the injured jumpers and those guarding the prisoners on the Motolance barge.

Immediately prior to initiating the takedown of the enemy positions Col. Randal pulled Ens. Slade aside, "The purpose of the exercise is to kill or capture the Nazis and not take casualties while we're doing it. Do you understand the difference between 'Fire and Maneuver' verses 'Fire and Movement'?"

Ens. Slade said, "Not exactly, sir."

Down on one knee using the needle-sharp point of his Fairbairn Fighting Knife to draw in the dirt, Col. Randal said, "Fire and Maneuver is when you have a fire support element and a maneuver element. Your fire support element remains stationary in order to lay down a base of suppressive fire while the maneuver element flanks or envelops the enemy in order to put them at a disadvantage.

"Do you know the difference between flanking and enveloping?"

Ens. Slade said, "Negative, sir."

Col. Randal said, "We'll get around to that later—with Fire and Movement you have your fire support element lay down a base of covering fire while the movement element advances on the objective and either overruns it or places demolition charges to blow it up.

"Do you read?"

"Roger sir, loud and clear."

"So, which one are you going to use right now?"

"Fire and Movement, sir."

"All right then, do it, Ensign."

Having his team gather around Ens. Slade said, "Colonel, you, Captain Jaxx, Mr. Treywick and Lieutenant Hamilton are the support element. On my command concentrate your fire on the windows of the building to our immediate front. Lieutenant Hamilton, try to get those 45mm Brixia rounds in all four of the windows as quickly as possible."

Lt. Hamilton said, "Hey, Presto!"

Col. Randal said, "Wilco."

Ens. Slade was off to a good start, taking charge and giving clear concise orders. When he walked away for last-minute coordination with his pair of Det 3 operators, Lt. Hamilton said, "That was impressive . . . teaching a tactics class in the middle of an operation, sir.

"Now Ensign Slade can explain Fire and Maneuver versus Fire and Movement to his people and sound like a tactical mastermind, sir. That was your plan—to make him look good?"

Col. Randal said, "Put those 45mm mortar rounds where they're supposed to go, Lieutenant, and we won't need any tactical masterminds."

Private Nathan Boatwright walked over armed with an OSS issue 9mm Model 42 Marlin aka "Avenger" submachine gun. "Ensign Slade attached me to your fire support team, sir."

Col. Randal said, "Take up a position near me, Boatwright. Short crisp bursts through the windows. The idea is to kill the Nazis if we can or keep their heads down while the demo team moves into position—either way works."

"Yes, sir."

Col. Randal said, "Hold fast while I run my magazine. Then Captain Jaxx will empty his. He'll be followed by Mr. Treywick with his twelve gauge. You commence when he's done. Then I'll open again—we're a relay. While we're engaged in that, Lieutenant Hamilton will be putting 45mm Brixia rounds through the windows. We want continuous suppressive fire while the demo team works their way to the building to place their charges.

"Be alert for any Germans moving to the second story. If they do, sing out. We'll shift our fires—clear?"

"Yes, sir."

"How do you like the Marlin, Boatwright?"

"It's a fine weapon, Colonel. Lighter than a Thompson. More accurate on full auto—rifle type stock, pistol grip forearm. . ."

Col. Randal said, "I'll be using this one for the first time. Anything I need to know?"

He was now armed with a 9mm Model 42 Marlin submachine gun recovered from one of the KIA Det 3 Operational Swimmers—not aware which man.

Pvt. Boatwright said, "Aim it from the shoulder like a carbine, sir. Front sight, squeeze—it's very controllable on full auto. Cyclic rate's about 150 rounds per minute faster than the Beretta M-38 I've noticed you customarily carry."

Ens. Slade gave the command, "Fire Support Team—COMMENCE FIRE!"

The Great Teddy may have been an illusionist, which he claimed was different from a magician, but the way he zipped a 45mm Brixia round through each of the windows as fast as he could reload was no illusion and it was not magic. He was fast and smooth as if he was performing, putting on one of his shows. He transitioned from window to window. The result was a steady muffled *BOOOOOM! BOOOOOM! BOOOOOM! BOOOOOM!* in rapid succession.

Col. Randal, Capt. Jaxx, Waldo and Pvt. Boatwright quickly fell into a rhythm, first putting a burst of four or five rounds in the bottom right window, then quickly shifting to the left. Sheltering in that building was proving to be an unhealthy choice for the 1st Mountain Division Jägers.

The continuous suppressive fire was working to effect. There was no return fire as Ens. Slade and his Frog teammate low-crawled to the building, each of them coming in from different corners—on the outside. They were advancing at 45-degree angles cognizant of staying out of the line of fire of the support element which was laying down a continuous stream of suppressive rounds. Once in position both Frogs produced a half-pound block of C-3 with a four-second fuse and a ring lighter attached.

Ens. Slade shouted, “Fire in the hole!”

That was the signal for both demo men to pull the rings on the fuse lighters and toss their blocks of explosives through the windows. It was also the signal for Lt. Hamilton to shift his 45mm Brixia shoulder fired mortar rounds to concentrate on the second-floor windows. Col. Randal, Capt. Jaxx, Waldo and Pvt. Boatwright maintained a constant stream of 9mm rounds through the ground floor windows while raising their impact to well over the demo team’s heads. They did not want the Germans throwing the explosive charges back out.

Col. Randal was impressed with how well the Marlin SMG handled. The forward pistol grip—unique to Allied war production submachine guns, was a substantial aid to staying on target. Pvt. Boatwright had not exaggerated the weapon’s merits.

The instant the charges exploded, blowing smoke and dust out the windows, the two demolitions men jumped up and dived inside firing their SMGs as they went. Col. Randal’s support element leaped to their feet and charged the building following in after them. At this point, closing with the enemy, speed and violence of action were of paramount importance.

Inside, seeing and breathing was difficult. It took a moment but eventually three crumpled bodies were discovered. Ens. Slade and his Frog demo teammate raced upstairs to clear the second floor. When they came back down Waldo offered him one of his long thin cigars, “So Ensign, how you likin’ duty in Raidin’ Forces so far?”

Ens. Slade was as pleased with the cigar as if he had been decorated for valor. “Has its ups and downs, Mr. Treywick.”

Capt. Jaxx said, “Same for the Nazis on this island—they’ve had a rough day at the office.”

Col. Randal said, “So Jack, where exactly *is* the German detachment that’s being relieved?”

LIEUTENANT GENERAL “GERONIMO” JOE MCKOY LED SUB-Lieutenant Peter Grey and the team of OSS OG Det 3 Operational Swimmers ashore. He did not land at the pier in the event the Germans had guards posted on it. They usually did. The Criminals were nothing if not predictable in how they went about an occupation tour. Typically the Noncommissioned Officer in Charge (NCOIC) would pick two goldbricks for the assignment, which was probably a difficult decision considering his troops consisted of men let out of prison to serve in a “penal unit.” Since the guard post was a boring semi-punishment detail where nothing ever happened, the men manning it had a bad attitude. Which did nothing to enhance their performance maintaining security.

The team landed at a small beach a quarter of a mile from the dock. Two Frogs were pulling the Mark VI depth charge along the water’s edge on the improvised sled constructed out of a pair of the High Speed Launch’s twelve-foot boathooks. When the patrol came to where they could see the pier in the distance, Lt. Gen. McKoy called a halt. He handed his 9mm Beretta M-38 submachine gun to Petty Officer 3rd Class Joe Nelson.

“Hold in place, Nelson, and wait for my signal. When you hear me clap my hands three times, move up off the beach and join me at the end of the pier. I’ll turn my red-filtered flashlight on.”

PO3 Nelson wondered what this was about but said, “Roger that, sir.”

Lt. Gen. McKoy disappeared into the night and headed straight inland off the beach. He looped around and approached the pier from the land side making no more sound than the shadow of a cloud passing across the moon. Two Germans were sitting on the edge of a sandbag checkpoint dangling their legs over the side staring out to sea smoking cigarettes. They were feeling sorry for themselves for missing all the action at the bar tonight—the 999th knew how to party.

They never heard him coming.

Like Captain Billy Jack Jaxx, Lt. Gen. McKoy rotated between a suppressed .22 High Standard Military Model D and his suppressed .22 Colt Woodsman. Tonight he was armed with the Colt—a handgun he had used in his Wild West shows for years. He stepped out of the night from the land side of the checkpoint and shot both Nazis in the back of the head point-blank. *WHIIIIIICH, WHIIIIIICH!* They went down without a sound. The Criminals' cigarettes rolled along the dock glowing in the dark.

Lt. Gen. McKoy adhered to Colonel John Randal's unwritten but well-established policy for .22 caliber weapons: "One to the head, two to the body." Or vice versa.

WHIIIIIICH! WHIIIIIICH! WHIIIIIICH! WHIIIIIICH!

Why take a chance? Also a Raiding Forces policy. He knew from long experience in a number of wars and incursions it never hurt to follow unit tactical SOPs. They were in place for a reason.

PO3 Nelson heard the three claps and saw the red light. "Move out, boys."

On arriving at the dock the Frogs found Lt. Gen. McKoy searching the two dead Criminals. Neither had anything of intelligence value. He was not expecting them to.

When the team came up, Lt. Gen. McKoy ordered, "Drag these Nazis behind that shed over there and dump 'em in the shadows. If any overzealous NCO shows up to inspect the guards, which ain't likely, he'll probably just think these two abandoned their post and went AWOL to visit their girlfriends. The 999th probably has a lot a' that goin' on."

"Yes, sir."

Lt. Gen. McKoy said, "Nelson, pull your depth charge around there too and stand by. Lieutenant Grey and I are goin' to go get us a guide. Then we'll see what you boys can teach me about blowin' up bars."

"Yes sir, General."

Lt. Gen. McKoy and S/Lt. Grey started walking up the cobblestone street. A mistake. They had not made it half a block before five men appeared out of the shadows and confronted them with evil-looking knives. Greeks out looking for stray Nazis. The fishermen did not seem to understand they were violating curfew.

S/Lt. Grey said, "They believe us to be Jerries, sir."

Lt. Gen. McKoy said, "I don't guess it'd do any good to point out they're not supposed to be out this time a' night."

The white-hot hatred Greek islanders throughout the Aegean felt for the Germans could not be fully comprehended by anyone who never suffered through occupation by low-grade rear-echelon troops belonging to a penal unit like the 999th —whose troops were classified as "Criminals", with little to no supervision. Lt. Gen. McKoy thought it interesting to note the Division's soldiers were held in such low esteem by the German High Command that they were not authorized to wear the German Eagle on their uniforms.

Considering the 999th's troops never felt constrained to obey the laws of society during peacetime, they certainly could not be expected to observe the niceties of international rules of warfare. Especially since the Division was habitually deployed by German Twelfth Army, the controlling military authority in Greece, to use terror as a weapon against the civilian population. The idea being to cow the citizenry into compliance with their occupiers. The Division had a license to murder and rape.

Being mistaken for a German by a mob of Greeks on a remote island in the Aegean in the dark of night was, "Not good."

S/Lt. Grey said something to the men in Greek. The response was a menacing rumbling that sounded like cursing, threats or both. The men had their courage up now, ready to swarm them with their knives.

Whatever he said, it did not have the desired effect.

"HALT!"

PO3 Nelson stepped out of the dark accompanied by two of his Frogs. They had their 9mm Marlin "Avenger" submachine guns leveled. OSS OG Det 3 may not have had much infantry training but they were some of the best in the business at short-range silent movement which is different from patrolling—a skill set useful for clandestine nocturnal beach surveys that might require a sentry to be eliminated.

Lt. Gen. McKoy said, "Try splainin' it to 'em again, Lieutenant."

The knives disappeared. The fishermen looked sheepish. Maybe a little disappointed. They agreed to lead the way to the 999th's bar.

Time was short. It was imperative the diversion take place, and right now. The patrol moved out but not before Lt. Gen. McKoy conducted a quick reorganization. The two Frogs dragging the depth charge on the improvised sled were replaced by the two biggest fishermen.

PO3 Nelson said, "That should teach 'em to pull knives on the General."

The Frogs did not respond other than to laugh but they thought it was a nice touch. From the moment Lt. Gen. McKoy took charge a bonding process had been taking place. Serious doubts initially existed among the Operational Swimmers because of the age difference, his being a lieutenant general, and the Wild Bill Hickock-length silver hair hanging out the back of his flat-brimmed bush hat. Now in record time the General was fully established as their leader based on demonstrated ability. The age differential no longer seemed a problem. The Operational Swimmers were now recognizing it for what it was… experience. He knew what he was doing.

They liked being one of his "boys."

OSS OG Det 3's confidence level the mission would be brought to a successful conclusion was growing by the minute. Even though they were on a remote enemy-occupied island, making a movement to contact, with the intention to attack an enemy objective where they would be substantially outnumbered—most likely at least three-to-one. On first hearing the assignment the plan sounded a lot like a suicide mission. However, tonight was the reason most of them had volunteered for the OSS—a high-value target of strategic importance to the U.S. war effort. While that might be a stretch in this case it was at least of interest to the top leaders of the United States and the United Kingdom—which to the Frogs was a big deal.

Some of the Operational Swimmers were beginning to come around to the belief small-scale clandestine ops under cover of darkness of the type Raiding Forces typically conducted beat the idea of clearing amphibious landing sites of obstacles off a beach somewhere in the Pacific in broad daylight in full sight of its Jap defenders.

S/Lt. Grey moved out with the Greek leading the way. As they traveled, he explained to Lt. Gen. McKoy the fisherman had informed him the place they were looking for was not more than three blocks away off Sikinos' main road just up ahead.

Lt. Gen. McKoy said, "Good to know."

The patrol turned off onto a side street not much wider than a cobblestone walkway. In the distance, bright light was coming from a building up ahead. The 999th did not practice blackout discipline when on occupation duty. Who was going to bomb a bar?

Lt. Gen. McKoy halted the patrol. The Frogs immediately went into a tight perimeter defense around the Mark VI depth charge. They went down on one knee providing 360-degree security.

The General knelt in the center and whispered loud enough for everyone to hear him, "PO Nelson and I are goin' to perform a leader's recon. Corporal Davenport, you're in charge here. If we're not back in ten minutes turn the Greeks loose and light this thing off right here in the middle of the road. Then hightail it back to the HSL and shove off."

Cpl. Davenport said, "Wilco."

Lt. Gen. McKoy said, "Synchronize watches—in thirty seconds the time is 2110 hours—hack."

Then he and PO3 Nelson moved out silently. They stayed to the side of the road where the buildings were casting shadows from the three-quarters full moon. It had been dark long enough that the Germans should not be wandering around outside. The 999th might be lax in implementing basic security measures but they were not crazy. Later in the night after they were blind drunk one or two might stagger outside alone for some unknown reason that only made sense to them at the time, but it was too early for any of the Criminals to be that far gone.

The bar was located on the corner of another cobblestone path. As they approached, raucous music could be heard blaring and people laughing. The party was clearly in full swing. Lt. Gen. McKoy and PO3 Nelson stayed to the shadows but edged up to where they could see in the plate glass windows. What was taking place inside came as a shock. Even to a man as experienced as Geronimo Joe.

The bar was crammed with Germans. More than had been anticipated. There were quite a few Greek women present, possibly from the Communist community—meaning they were disposable. Everyone seemed to be enjoying the night's floor show.

In the center of the room was a table with a woman dancing on it—or so it seemed at first glance. Closer inspection revealed the girl had her hands tied behind her back and there was a wire noose around her neck looped over one of the ceiling beams. It was strung tight. And she was not dancing on the table. She was standing on her tiptoes on top of a block of ice set on the table.

It was melting.

The drunken crowd of men and women gathered around watching seemed to think her predicament was hilarious. Like the pictures of a jeering mob at the guillotine during the French Revolution. The kind you saw in world history schoolbooks during junior high school and tended to doubt actually happened.

Lt. Gen. McKoy said under his breath, "That's some sick sons a' bitches in there."

PO3 Nelson could not think of a response. Not even a one-word answer. Nothing in his life experience or the intensive training he had gone through to be an Office of Strategic Services Martine Unit Operational Swimmer prepared him for anything like this.

Lt. Gen. McKoy said, "I've seen enough."

They made their way back to the patrol. Everyone in the perimeter pulled in tighter to better hear the frag order they knew was coming. The Frogs were hyped up—ready to get the show on the road.

Lt. Gen. McKoy had a word with PO3 Nelson. "Do you have a blasting machine?"

PO3 Nelson said, "Roger that, sir—always."

Lt. Gen. McKoy said, "OK, forget the ring main system idea. I want to place the depth charge against the bar's front door and touch it off with a command detonation. What I don't want is a burnin' fuse we have to wait on to blow. I say the word, I want instantaneous initiation—boom!"

PO3 Nelson said, "No problem, can do, General."

The Frogs were experts in the use of all types of military explosives and what Lt. Gen. McKoy wanted now was a simple device to rig—easier than the ring main system.

Detonation cord is generally ignited by a Switch No. 10 Time Pencil. British nomenclature—they invented it for Special Operations Executive and the time pencil was subsequently adopted by the Office of Strategic Services, though it also had the U.S., Lighter, Fuse, Weatherproof, M2 in inventory for non-electrical blasting caps. When activated the time pencil is essentially a burning fuse contained in a waterproof tube—although that is not exactly how it works—set with a color coded delay from ten minutes to twenty-four hours and used to detonate non-electrical explosives like detonation cord. The idea for the delay is to give the demolitions man time to depart the area. Det cord looks like a flexible nylon tube but it is filled with pentaerythritol tetranitrate (PETN). When detonated the cord/tube appears to explode all at once—very fast. The cord sets off the non-electrical blasting cap which in turn ignites the charge it is attached to. Det cord is handy for all kinds of military uses but you do not have instantaneous detonation.

With a Mark 27 Navy/Marine firing key, like the one PO3 Nelson carried, generally referred to as a "blasting machine," a current is sent along the length of electrical wire attached to an electrical blasting cap that is used to initiate the explosive device—in this case a 300-pound Mark VI depth charge. There is an almost simultaneous explosion. Exactly what Lt. Gen. McKoy wanted. He was not interested in waiting ten minutes for it to blow up.

Nevertheless the General wished he could wrap a loop or two of det cord around every lowlife in the bar and light it off at the same time.

He said, "Listen up boys, we're not goin' with the ring main system as previously briefed. What's about to happen is we'll roll our depth charge up against the bar's front door, run electrical wire back to where PO Nelson can hook it up to his blasting machine and blow that Nazi hellhole to kingdom come."

Though they maintained noise discipline it was clear the Frogs liked the sound of that. Now they were finally doing something they were trained for. This mission was getting better and better.

Lt. Gen. McKoy said, "Once it blows I'm goin' to make entry and shoot everyone in the place. Any a' you harborin' concerns about man's inhumanity to man—meaning me killin' bad guys, get yourself to the front a' the column when we move out. You'll see somethin' that'll be givin' you nightmares the rest a' your natural life and it won't be me shootin' Nazis and their girlfriends in the head."

PO3 Nelson said, "The General ain't exaggerating—it's ugly up there."

Lt. Gen. McKoy said, "You boys'll never have yourself another problem dropping the hammer on the opposition ever again in our AO—men *or* women."

No one broke the silence but the Frogs wondered what had happened to cause the outburst—something changed. Lt. Gen. McKoy had always been the image of laid-back. They broke out det cord from their demo kits and started splicing it together. PO3 Nelson inspected his Navy/Marine Corps MK 27 Firing Key. It was capable of handling up to ten blasting caps at a time but tonight they would only need one. Then he inspected his men's splicing.

"Good to go, General."

Lt. Gen. McKoy said, "Lieutenant Grey, you stay here with the Greeks. As soon as you hear the explosion, release 'em. Then come link up with us."

"Wilco, sir."

PO3 Nelson said, "How do we transport the depth charge from here to the bar, sir? Dragging it will make noise."

Lt. Gen. McKoy said, "They're designed to roll off a rack—roll it down the street."

This order was met by disbelief from the Frogs. Sounded insane. They had not been in Raiding Forces long enough to have Waldo

Treywick explain Occam's razor. "The simplest solution is best." Which never-the-less would have sounded crazy this night.

PO3 Nelson designated two of his people to be the rollers. The men felt foolish knowing they were about to violate every principle of clandestine security in the book. Never-the-less when the whispered command to "Move Out" came, the two Frogs started rolling. Out in the open. Right down the center of the street. In the middle of an enemy-occupied village. Were they really doing this? On reflection, rolling a depth charge up to the front door of a bar full of drunk Germans was not any crazier than anything else that had happened today.

Which was not all that reassuring.

THE REST OF THE TEAM MOVED ALONG THE EDGE OF THE wall staying to the shadows with their 9mm Marlin "Avenger" submachine guns at the ready, praying the Frogs rolling the depth charge would not be spotted. The drawback to being in a brightly lighted room like the bar at night—you are unable to see out. Not that the revelers would be spending much time looking out the windows when they could be watching a naked woman standing on a melting block of ice trying to delay hanging herself.

There was no way to save the woman. Lt. Gen. McKoy was not even going to try. He had a long-standing personal policy of not risking live troops under his command on lost causes without serious extenuating circumstances.

As the depth charge rolled along it was making a metallic crunching sound like a car driving on a rim. To the Frogs doing the rolling it felt like the whole Nazi army could hear them coming. As they neared the bar the door opened and a German stepped outside. He glanced up the road and spotted the two rolling the depth charge his way. But having impaired night vision due to coming out of a lighted room into the dark, he was not able to make out the details of what was taking place.

One quick-thinking OSS OG Det 3 operator raised his hand and waved. The Criminal flicked his cigarette away and waved back. Could he have thought there was another keg of beer on the way? No way to know. The German went back inside.

As the depth charge rolled past the front of the file of Frogs, Lt. Gen. McKoy stepped out and fell in behind. "Pretty sharp thinkin' there, Malinowski."

As they approached the bar the Frogs rolling the depth charge turned to him for instructions. He pointed at the front door. They rolled the depth charge up tight against it. Lt. Gen. McKoy took a pair of Colt .38 Super magazines and wedged them under the rims. Now it was not going anywhere. The people inside the bar were trapped.

While he was doing that the two OSS OG Det 3 operators had a chance to peek in the windows and see what was taking place inside the bar. They froze. "Expect the Unexpected" was Rule No. 7 of Raiding Forces Rules for Raiding which the Frogs had to memorize first day. The men also were aware Col. Randal had retired it because in practice the unexpected had proven impossible to anticipate. But he never took it off the list of his Standing Orders.

Now, seeing the woman on the block of ice desperately trying to maintain her balance as it melted away while the drunken mob taunted her, the Frogs understood why Rule No. 7 had been mothballed. No one could have foreseen something like this—ever.

As they withdrew, PO3 Nelson moved forward to attach the blasting cap. Then he spooled out the electrical wire back to where the patrol was waiting across the side street around the corner. The petty officer had wisely kept the blasting machine with him in full compliance with another of Raiding Forces' unwritten policies: "Why take a chance?" It was also standard SOP demolitions best practice that had been hammered into the Frogs from day one of their UDT training.

When handling explosives accidents can and do happen.

As the petty officer was hooking up the electrical wire to the blasting machine, Lt. Gen. McKoy said, "I'm goin' to take your people around the corner for a minute. Give 'em a chance to let everybody see for themselves what they're dealin' with here. When I move 'em back you be standin' tall ready to initiate on my command."

PO3 Nelson said, "We need to have everyone around the side of the corner first, General. When the depth charge blows you're going to get your big bang—the equivalent of an airstrike. The book says airstrikes are not supposed to be called in within one mile of friendly troops. That's excessive, probably a holdover from antiquated prewar safety measure, but we need to be shielded from the shrapnel effect and the concussion.

"That last is going to be pretty much impossible up this close, sir."

Lt. Gen. McKoy said, "Service in Raidin' Forces is not without its vicissitudes."

PO3 Nelson said, "Don't we know it, General."

Lt. Gen. McKoy had the patrol pull into a tight perimeter again. "I'm taking you boys on a sneak and peek. Follow me around the corner. Take a quick look-see through the big plate glass windows. Then we're comin' right back and it's on."

There was a murmur from Frogs. They liked the idea of being allowed to see for themselves. The General was turning out to be an exceptionally good patrol leader. Not only did the former Arizona Ranger/US Marshall lead from the front and set the example—he made an effort to keep them current on developments.

They liked that.

Lt. Gen. McKoy said, "I'll lead the file. Everybody get yourself a good look. When I say 'take it home' do an about-face and head back here pronto. The big show'll be ready to start.

"Prepare to move out . . . move out."

Lt. Gen. McKoy led the way, staying to the shadows. When they filed around the corner far enough to be opposite the big plate glass windows he came to a halt. The men could clearly observe what was taking place inside. Trained to be stealthy in immediate proximity to the enemy, nevertheless the Frogs could not help gasping, with some not-so-quiet muttered cussing. It was a scene out of Dante's *Inferno* in technicolor. The naked woman was desperately trying to maintain her balance on the melting block of ice while her neck was being stretched by the wire noose—she was screaming. The crowd was laughing.

The Frogs were hard men. One of them had tears running down his cheeks. No one thought any less of him for it.

"Take 'er home, boys."

The team filed past where PO3 Nelson was kneeling with the Navy/Marine Corps MK 27 Firing Key at the ready and took up a position against the wall. Highly experienced with demolitions, the Frogs knew what was coming next even though no one in OSS OG Det 3 had ever been this close to a three-hundred-pound charge designed to destroy enemy submarines underwater when it detonated. It was going be loud.

Assessing the situation and having noted how observing the scene inside the bar had affected the Frogs, Lt. Gen. McKoy said, "Change a' plans, boys. When the charge goes off and we're sure we're all still alive walkin' and talkin', I want you to double time back around to where we were. Everybody fire off a magazine's worth a' 9mm through those Marlin guns a' yours aimed at where the plate glass windows used to be—then check fire and standby holdin' in place. I'll go over and clear what's left a' the bar."

There was nothing he could have said to them that would have made the OSS OG Det 3 Operational Swimmers much happier. The men were shaken by what they had seen. Nothing had prepared them for anything like the casual brutality of the war in the Aegean. Lt. Gen. McKoy was right—the Frogs were never going to have a problem pulling the trigger after tonight no matter the circumstances. Not one of the extensive "Enemy Forces" briefings during their two years of training readied them for the barbarity they had witnessed.

Not even close.

And Lt. Gen. McKoy was dead on about something else but not mentioning it to anyone. The Frogs were never going to be the same. They were damaged goods.

"Light it off, Nelson."

KEEEEERBOOOOOOM! The explosion was a blockbuster magnitudes louder than expected. It was like standing next to a volcano that erupted during an earthquake that registered an 8 on the Richter scale. Not only was the sound beyond description, the wave of pressure created by the concussion body slammed the team. The men felt like their chest cavities were about to cave in and they were having a difficult time trying to breathe. Everyone knew to keep their jaws unlocked but it did not do all that much to help. And while they held their ears they were still ringing. The entire patrol was experiencing headaches, their vision was blurred and their equilibrium discombobulated.

Lt. Gen. McKoy got his big bang, but the team paid a price. “Move out, boys.”

The Frogs staggered around the corner. When they came on line opposite the semi-destroyed bar, the men opened fire on what was left of the building. The corner was gone—vaporized. Two walls had almost completely disappeared. Part of the ceiling was down. Water was pouring from a broken pipe. The inside of the place was reduced to rubble. However, the beam supporting the wire noose had remained largely intact though part of it had jarred loose from the roof and was unsupported. It was sticking out at an angle.

The woman was swinging back and forth only inches off the floor . . . with any luck dead before she strangled.

While the firing was ragged—several of the Frogs were unsteady, having trouble focusing their eyes—the exercise served its intended purpose. It let the troops feel like they were hitting back individually. And that was all Lt. Gen. McKoy wanted. After the men emptied their 25-round magazines, he said, “Well Petty Officer you boys sure lived up to your reputation this night—blew that place to smithereens.”

None of the OSS OG Det 3 Operational Swimmers would have wanted anyone to know how good the praise made them feel.

PO3 Nelson said, “The P for Plenty Formula sure delivered tonight, sir.”

Lt. Gen. McKoy said, “Ain’t no such thing as overkill.”

Then he walked across the street, ducked under the partially fallen ceiling toward the front of the room, went around and shot every person inside—alive, dead, conscious, unconscious, man or woman, exactly like he said he was going to. All twenty-seven of them. Most were tightly clustered in a circle under the beam where the woman was swinging.

PO3 Nelson waited until Lt. Gen. McKoy called "Clear," then brought the Frogs over to inspect the damage. The men stood looking at the mass of dead bodies not saying a word trying to avoid staring at the dead woman dangling from the beam. The emotion they felt was . . . satisfaction.

The OSS OG Det 3 Operational Swimmers wondered what the woman had done to deserve such a humiliating end. Then it began to dawn on them—maybe nothing. She was simply the night's entertainment.

Lt. Gen. McKoy said, "Crowd got a' little more show than they bargained for. This is our enemy, boys. Don't forget it."

Corporal Raymond Davenport said, "Karma's a bitch."

Roger that.

DR. LAYTON WINTHROP HAD A MULE FOR FLYING OFFICER Bash Potgieter. Finally seizing one at pistol point after the third farmer they called on refused his request. Normally mild-mannered, the professor was on the point of shooting the man when the Communist finally agreed to the loan of a mule. Provided it was turned loose after they were finished. He wanted to be able to say the animal had been stolen if the Germans showed up.

Not a problem.

The flying officer was hoisted on the mules back and they were moving toward the rendezvous point when a rolling *BOOOOOM*! was heard in the distance from the direction of the village. Dr. Winthrop was impressed. If that was Lieutenant General "Geronimo" Joe McKoy's diversion it was a masterpiece of epic proportions.

Now the team was moving at speed. The professor was proving to be a gifted compass man. They arrived at the extraction point in the exact spot the LCR(S) was pulled ashore. A feat of land navigation over rugged terrain without the benefit of a contour map only an experienced land navigator who has spent hours and hours stumbling around in the dark of night perfecting his tradecraft could appreciate. Not having to search up and down the beach to locate the boat, as frequently happened on exfils, was a valuable time saver. And it reduced the stress level associated with the possibility of not finding it before daylight.

Even the Lovat Scouts were impressed. They appreciated fieldcraft.

Private First Class Norvel "Horn Dog" Hansen said, "That right there was some good navigation, Doc."

The mule was released as promised and everyone loaded into the LCR(S). The Life Boat Service Men were raring to go. They attempted to set the record for a two-mile open water paddle. Had there been such a record the LBSM might have broken it.

PFC Hansen radioed Long Range Desert Patrol Baker Team 6 the Command Party was en route. Since the LRDG radio operator had been monitoring the professor's network the patrolmen already knew the team had rescued the pilot. When the LCR(S) arrived, the former desert reconnaissance specialists, now long-range beach watch operators who insisted on being called "patrolmen," were packed up and ready to go. Their gear was already stowed onboard the Walrus. Wing Commander Paddy Wilcox had the motor ticking over.

FO Potgieter was loaded and Dr Winthrop's men scrambled in followed by Baker Team 6. After a short takeoff run, the plane was airborne and on the way back to ABC. It would have been hard to imagine a more perfect mission. It did not hurt that they had been going against the Criminals of the 999th Light Division. Nevertheless, it was a beautiful operation executed with "surgical precision."

Which only happens in the movies.

LIEUTENANT GENERAL "GERONIMO" JOE MCKOY LED HIS OSS OG Det 3 team of Frogmen back to the beach where the HSL was waiting just offshore. They took their time as they went firing their weapons and trying not to hit anything. The idea was to create additional confusion for any Germans who might not have been in the bar when it blew up. The OSS OG Det 3 operators and Lt. Gen. McKoy were enjoying themselves rebel yelling and blazing away. Shooting up the town was a good way to let off steam and try to forget the part of the night the Frogs were never going to forget. Something the men would find out soon enough.

The General already knew that.

Before going onboard the HSL, he told Sub-Lieutenant Peter Grey, "Get with Lieutenant Featherstonehaugh. These people are going to go below and rack out. When they wake up I want them to be at Advanced Base Castelrozzo."

S/Lt. Grey was still suffering shock from the depth charge blast and what he had witnessed in the bar, "We can arrange that, General."

CAPTAIN BILLY JACK JAXX ORDERED THE FROGS TO GIVE him all the 1st Mountain Division knives they had recovered. He had seventeen of them strapped on his body or in his pockets. They had fourteen more. . . for a total of thirty-one of the unique little stag horn handled blades. Jack Cool had plans for them. They went into the pack Lieutenant Ted Hamilton had virtually emptied of 45mm mortar rounds for safekeeping.

Capt. Jaxx said, "Make sure you get home with them, Lieutenant."

"Sir!"

Colonel John Randal ordered the six surviving Germans be assembled on the Motolance barge. Then he ordered Lieutenant Chase Starrett to assess the captured weapons for any that might be of use to Raiding Forces. There were two 9mm PO8 Lugers but the rest were of no interest. The police chief and two of his men turned up now that the firing had ceased. They were presented with the captured weapons. The chief was instructed to take charge of the Nazis. What they did with them was their business. Col. Randal had a policy of not interfering with internal Greek affairs once a mission was completed.

Lt. Hamilton fired off the red and green flares to signal Beverly she was cleared to land. The Hudson came in, touched down and taxied to the pier. To the hard-used Frogs, some of whom were so banged up from the pair of jumps in one day they were barely capable of walking unassisted, the ungainly aircraft was a beautiful sight. One they never thought they would live to see.

When the plane pulled up to the dock, the men of OSS OG Det 3 cheered.

Col. Randal ordered, "Saddle up—let's get the hell out of Dodge."

The Greeks were shooting the prisoners before the first Frog made it onboard the aircraft.

BECAUSE THE HUDSON WAS FASTER THAN THE WALRUS the two aircraft arrived at Advanced Base Castelrozzo within minutes of each other even though they had taken off at different times and the Hudson had a longer flight. The planes landed shortly after 2400 hours. Everyone in Raiding Forces and most of the other Allied personnel who had small detachments on ABC—MI-6, SOE, MI-9, PWE, A-Force, etc.—turned out to welcome Captain Billy Jack Jaxx home. The fact that he was on the *Perseus* when it had been declared missing had not been kept secret for long. Such a tiny island—inhabited by representatives of more intelligence agencies per square foot than most likely any other place on the planet—made that impossible.

The result was entirely predictable. Capt. Jaxx was a popular officer, especially among the female contingent of Royal Marines, FANYS, WRENs and an impressive number of local Greek women. They had been in mourning ever since the news leaked he was MIA—presumed dead. Major the Lady Jane Seaborn locked herself in her suite after giving Flanigan strict orders she was not to be disturbed. The former policeman risked violating those instructions to report the signal Beverly

Blackwell radioed that Colonel John Randal had jumped on Hydra leading an attempt to retrieve Capt. Jaxx who might or might not be at large on the island. The information did nothing to improve her morale since Vice Admiral Sir Randolph "Razor" Ransom had explained privately it was unlikely for anyone to have made it off the submarine alive. And that he believed the radio message claiming to be from Capt. Jaxx was a trap.

Lady Jane had been crying for hours.

The news that Capt. Jaxx and Flying Officer Bash Potgieter had both been rescued and were en route back to base was met with equal amounts of relief, joy and disbelief. The odds against such an outcome were astronomical. Raiding Forces had an unofficial motto based on a slight variation of a quote from a nineteenth-century French novelist Lieutenant Colonel Sir Terry "Zorro" Stone had engraved on a brass desk plate: "The Difficult We Do Right Away . . . The Impossible Takes A Little Longer."

Around ABCHQ tonight's operation was already being called "The Magnificent Mission."

When the planes came in, the abbot of the local Greek Orthodox Church ordered the ringing of its bell to celebrate—how he was aware Capt. Jaxx was on board or that he even knew him was an open question. Someone was shooting off parachute flares turning night into day. And a crowd was milling around on the dock.

As the Hudson taxied to the pier Col. Randal was sitting in the co-pilot's seat next to Beverly who pointed out, "I'm pretty sure that's Lady Jane front and center on the landing with Happy. My guess is you're about to be in big trouble."

"Maybe I should put my splint back on."

"Not enough to save you this time, Johnny."

Col. Randal said, "Yeah, roger that."

THE WALRUS DOCKED FIRST BECAUSE FLYING OFFICER Bash Potgieter required immediate medical attention. Doctor Stephen Milam was on hand with a pair of stretcher bearers to transport the South African to the Aid Station. It would not do for the pilot to die at this stage after all the effort that had been expended to rescue him. The medical team went onboard as soon as the plane pulled in.

Doctor Layton Winthrop followed the stretcher off with the Lovat Scouts and Private First Class Norvel "Horn Dog" Hansen and then the Long Range Desert Group Beach Watch Team 6 patrolmen. They were met by cheering from the jubilant crowd. Everyone understood the men had pulled off one of the most difficult assignments, with the least likely chance of success, any team of special forces operators could be given.

The Hudson docked after the professor's people were ashore. Colonel John Randal assigned the order of march for disembarking. Captain Billy Jack Jaxx led out first, Beverly Blackwell second, Lieutenant Chase Starrett third, Lieutenant Ted Hamilton fourth, Ensign Westly Slade was fifth, followed by the men of OSS OG Det 3, some of whom required assistance. He was last to go ashore.

Almost. There were two dead OSS OG Det 3 Martine Unit Operational Swimmers who would be removed from the cabin later. The jump on Hydra resulted in a high casualty rate. A unit the size of Raiding Forces could not sustain those losses for any length of time and remain combat effective.

Jack's Pack mobbed him when he led the team off the plane. A number of the female Royal Marines, FANYs, WRENs and local Greek women had to be restrained by the Vulnerable Points Wing security specialists who were on hand in the event of just such a development. Emotions had been running high ever since word had leaked out Capt. Jaxx had gone missing and even though injured, Col. Randal was leading a mission to rescue him.

When he finally exited the aircraft the crowd went wild. Flash bulbs were popping. Happy raced along the pier and tried to jump on him in joy. Major the Lady Jane Seaborn could not restrain herself. She ran over and did jump in his arms, nearly knocking him down. She gave him a passionate kiss—much to the delight of the cheering crowd.

So much for no public displays of affection.

This all came as a shock. The euphoria of a successful mission had already worn off on the flight back. Col. Randal was having to live with the fact that two of his men were KIA. Command responsibility had never been a heavy burden for him but he took the lives of his troops seriously. He led the two Frogs in harm's way and they were not coming home. It was hard to celebrate knowing there were two blanket-covered bodies still on the Hudson waiting to be removed. And back in the States two families would be receiving black rimmed telegrams—the border made it unnecessary to read the message.

Lance Corporal Ray "Tank" Karlsson picked Col. Randal up and hoisted him on his shoulders. The Marine Raider/Frog carried him all the way up the path to ABCHQ with his teammates alongside pumping their weapons in the air in celebration. The cheering crowd followed. Lady Jane was laughing and crying at the same time. Happy was dancing around in a circle.

Randal said, "Tank, what the hell?"

LCpl. Karlsson said, "Hang tight, Colonel, anybody who'd jump in with us on a bum leg doesn't need to be climbing this hill. Besides you might want to make another one—the day ain't over yet."

No one not a badged member of Raiding Forces or staffing one of the Allied offices on the island was allowed inside the building. Still, the spacious room that served as the TOC was crowded. Before any of the operators received medical treatment a short debrief took place. It was always best to conduct them while the mission was fresh on all the participants' minds. A much more thorough finger-pointing session would be conducted later in semi-private in order to focus on mistakes made and lessons learned.

Capt. Jaxx went first. You could have heard a pin drop in the crowded Tactical Operations Center as he described his escape from the *Perseus*. In typical fashion, Jack Cool made it sound like another day at the office. Nevertheless, Lady Jane and Beverly were sobbing into tissues as were a number of the other service women. Even Mandy had tears in her eyes and crying was not her style. More than a few of the men present had dry throats. Volunteers for future submarine missions were going to be hard to come by from this crowd.

Protocol during debriefings—the speaker is never interrupted. But when Capt. Jaxx finished describing the swim ashore, Vice Admiral Sir Randolph "Razor" Ransom, a man rumored to have a "Heart of Oak' as the lyric went in the official marching song of the Royal Navy, could not constrain himself.

"I want to commend Captain Jaxx in the highest possible terms. Having never made a voyage on a submarine, he listened to his safety briefings, paid attention, knew exactly what to do in the event of needing to escape from a submerged sub in distress and in so doing—when put to the acid test—performed in accordance with the highest traditions of the Royal Navy. On my word, be assured the Captain shall be decorated for his extraordinary performance.

"My apologies, Jack. I had to inject that—continue your debrief."

Next Capt. Jaxx walked the audience through his actions on Hydra step by step up until the moment Col. Randal and the stick of OSS OG Det 3 dropped on the island. A hushed room grew even quieter minute by minute. The onlookers were hanging on every detail. There were no superheroes in Raiding Forces. One of the things constantly drilled in during training at the Commando Depot and repeated from time to time in the unit to this day was, "You are not Supermen . . . there is no 'S' on your chest." However, Jack Cool was fast approaching true action hero-type status.

When he finished, the men of OSS OG Det 3, as bruised and battered as they were, stood up and cheered. A British team returning from a mission would not have done something like that. Displays of emotion were rare at Raiding Forces' debriefings, understatement being the norm, but it was something the Frogs had been doing. The Operational Swimmers were impressed by Capt. Jaxx's one-man army act and wanted to show it.

Doctor Layton Winthrop went next. He described the mission to rescue Flying Officer Bash Potgieter from the time the Walrus arrived at the Long Range Desert Patrol Beach Watch Team 6's position. The professor made his party's actions sound so matter-of-fact as to be boring. No one in this crowd actually believed that.

"I shall be recommending Scouts Ferguson and Fenwick for the Military Cross and want to personally thank Colonel Randal for having the foresight to assign them to my team. Without the Lovats' exceptional fieldcraft and tactical acumen when carrying out the actual rescue it is doubtful our mission would have resulted in success."

Col. Randal went last. He walked the audience step by step through the Hydra operation from the time he received intelligence that Capt. Jaxx might be located on the island until the Hudson landed to bring his team home. Singled out for individual recognition were Beverly Blackwell, Lt. Hamilton and LCpl. Karlsson. The OSS OG Det 3 operators who jumped on Hydra were praised for "stepping up and continuing the mission even when fighting through injuries that would have sidelined lesser troops."

"As for Captain Jaxx, I don't even know where to start. . ."

The audience roared—there were a lot of relieved people in the room.

Col. Randal said, "Mandy, you have something for me?"

"Yes, I do." She walked to the front with a small velvet jewelry bag.

Col. Randal ordered, "Det 3 on your feet—assume the position of attention."

Taken off guard, the Frogmen struggled out of their chairs. What was this?

Col. Randal said, "In recognition of your making a forced entry parachute assault on an enemy-held island without the benefit of advanced intelligence on the hostile forces you would be encountering *and* dropping on a hot DZ, I am awarding your Parachute Wings prior to completion of the five jumps needed to qualify. You'll have to finish Airborne School. Can't have you Frogs jumping out of airplanes without training."

Everyone in the TOC laughed.

"I'm proud to have you men in my command."

Then Col. Randal worked his way down the rank taking his time pinning on each Operational Swimmer's silver parachute wings as Mandy handed them to him. "Congratulations . . . good job, stud."

When he came to Ens. Slade he said, "I'm recommending your promotion to Lieutenant junior grade effective immediately. You're my new OSS OG Det 3 commander. I've got plans for you and your people, Lieutenant."

"Yes sir . . . thank you, sir."

The ceremony completed, Lady Jane announced, "Det 3 report to Doctor Milam immediately. You shall find him standing by in the Aid Station. Once he has finished treating your injuries our traditional post-mission meal is waiting in the Other Ranks mess.

"Well done, lads."

Lady Jane and Happy followed Det 3 down the corridor to the Aid Station to make sure the men actually went to be checked out and to personally evaluate the extent of their injuries. They were, after all, *her* OSS OG Det 3 Frogmen. When they arrived the dog went around and made each Frog pet him.

While they were being treated the troops were telling jump stories nonstop. They were hilarious. Dr. Stephen Milam and his medical orderlies were rolling in laughter. Every Frog was doing his best to outdo the last one. The stories all centered around what happened once the order came to "chute up" onboard the Hudson, the chaotic exit with LCpl. Ray Karlsson aka "Tank" pushing them to the exit door like peas out of a pod while dragging one man—or the hair-raising PLFs in downtown Hydra Port on a DZ so hazardous one of the jumpers broke his neck and died.

The actual firefight with the 1st Mountain had not seemed to make much of an impression on anyone even though one of the team had been shot and killed in the air while landing. The stories were all centered around the drop. Except for the one where Col. Randal shot two Germans with his pistol who were in the act of running him through with their bayonets—that had gotten the Frogs' attention.

There was a lot of laughing about Col. Randal flicking open a switchblade knife onboard the C-47—a clock-stopping moment you probably had to be there to appreciate. Lady Jane wondered what that was about but chose not to ask. She marveled at being able to have fun and laugh while every single Frog received some form of medical treatment. Hilarity at a time like this did not seem like normal behavior but it definitely made everyone present feel better.

OSS OG Det 3 may have been reduced to walking wounded as a result of the Hydra mission coming on the heels of their earlier hard-luck jump on ABC. But the Frogs were exhibiting sky-high morale. Lady Jane noted this was a different group of men than the cocky egotistical Operational Swimmers who reported in from the States only days before. At long last the men of OSS OG Det 3 were no longer military nomads traveling endlessly from training station to training station.

The Frogmen had found a home in Raiding Forces and were making clear they thought it a good fit.

COLONEL JOHN RANDAL AND MAJOR THE LADY JANE Seaborn finally made it to their suite after a high-spirited post-mission meal in the Other Ranks mess. He gave strict orders to the Vulnerable Points Wing NCO on the duty desk outside the door they were not to be disturbed. No matter what short of another invasion of German Brandenburgers. .

"Sir!"

They went out to the private pool on the patio, ripped off their clothes and dived in. It had been a long day and the water felt good. Standing in the shallow end Lady Jane wrapped her arms around his neck staring him in the eyes.

"Simply had to play the hero. Do you have any idea how furious I was at you? What could you have possibly been thinking to make a low-level parachute jump with your injured leg knowing you were most likely dropping into a German trap?"

"Actually, I was thinking of you."

Col. Randal thought that was a pretty good answer but probably not good enough.

Lady Jane let it go by. "The Frogs came back changed men—no more attitude. Fabulous stories. Everyone had an anecdote about you. What happened on board the Hudson . . . sprinkle fairy dust on their swim flippers while they napped?"

"Works every time."

"General McKoy says when it comes to handling troops you have a magic touch."

Col. Randal said, "What might be your thoughts on that subject Lady Jane considering you're one of my troops?"

"I *know* you have a magic touch."

THE HIGH SPEED LAUNCH ARRIVED AT CASTELROZZO THE next morning. The entire contingent of Raiding Forces on the island and other attached personnel were on hand to greet it as they had been when Colonel John Randal's and Doctor Layton Winthrop's parties returned. Vice Admiral Sir Randolph "Razor" Ransom was on the dock with Col. Randal and Major the Lady Jane Seaborn to welcome back his Small Raids Inc. crew on the HSL returning from their maiden voyage supporting Raiding Forces.

Lieutenant General "Geronimo" Joe McKoy led his OSS OG Det 3 team ashore. While this was an exuberant moment, the Frogs were wearing frozen expressions—commonly described as the thousand-yard stare. Something had happened on Sikinos. That was immediately apparent to the people on the dock. The instant Lt. Gen. McKoy stepped ashore Col. Randal snapped to attention and rendered a salute.

A faint smile played around Lt. Gen. McKoy's lips as he returned it. These were not two men adhering to formal military protocol. This was respect publicly given and publicly returned.

Once in the Tactical Operations Center for the debriefing, Lt. Gen. McKoy did something surprising. "I'm going to let Petty Officer Nelson do the honors."

Petty Officer 3rd Class Joe Nelson was taken off guard but he stepped up to the front of the room and walked the audience through their mission starting from the time Lt. Gen. McKoy took charge of the team on the HSL. He explained how they had decided to use a depth charge to take out the German's tavern "to increase the lethality of our demolitions package and enhance the volume of the explosion incidental to the diversion."

Lt. Gen. McKoy whispered to Col. Randal, "I'll be nominating Nelson for our Officers Candidate Program, John."

"Good."

PO3 Nelson explained how Lt. Gen. McKoy eliminated the sentries. How five local Greeks armed with knives confronted him a few minutes later thinking he and S/Lt. Grey were Germans and how they were convinced of the advisability of putting their blades away and guiding the team to the bar where the 999th could be found. "Which they did with alacrity."

At that point, the Petty Officer looked to Lt. Gen. McKoy who said, "Go ahead—lay it out, Petty Officer."

PO3 Nelson said, "The General and I conducted a leader's recon. As we approached the bar we could see inside through the windows. What we saw will be scorched in my memory forever. There was a drunken mob of men and women taunting a naked woman on a table who had a wire noose around her neck tied off on an exposed beam in the ceiling. She was standing on a block of ice.

"It was melting."

A room full of highly experienced Special Operations veterans and long-service supporting staff went quiet. No one in the TOC was a stranger to Nazi atrocities. But for pure cold-blooded satanic evil this story was in another league. The enormity of the German 1st Mountain Division executing 6,000 Italian soldiers on Cephalonia was almost impossible to grasp, but one publicly humiliated woman, naked, standing on a melting block of ice, knowing she was about to die to gratify the pleasure of a drunken mob—that you could picture in your mind, almost. If you believed it was true.

"We could see that the woman was screaming but we could not hear any sound coming out."

At this stage of PO3 Nelson's rendition of events, people in the TOC were no longer making eye contact with each other—horror and disgust were trumped by embarrassment.

"Then we rolled the 300-pound depth charge up against the front door of the bar. General McKoy wedged it in place with a couple of his pistol's magazines so no one could escape. We ran det cord to our Mk 27 firing key and on command lit it off. The resulting explosion blew out the front of the building and knocked unconscious or killed everyone inside.

"The General then walked over to what was left of the place, went in and shot everyone to make sure they were dead—twenty-seven people by my count, men and women. When General McKoy gave the all-clear our entire team was brought in to examine the results. What took place in that bar was a crime committed against humanity by demented individuals who were taking gratification from torture, humiliation and murder. We killed them all, every last one and were glad of it.

"After that we commenced our exfil, firing our weapons as we withdrew. The intent was to add to the state of confusion on Sikinos caused by the explosion and add to the diversion to cover Doctor Winthrop's withdrawal. The HSL came in, picked us up and here we are."

A hush had fallen over the TOC. The war in the Aegean, never fought to Geneva Convention standards, had just taken a horrific turn. The mental image of the 999th Light Division's "Criminals" debasing a single Greek woman and enjoying it hit home harder than the report of the mass execution of 6,000 Italian soldiers.

Now it was clear beyond any reasonable doubt the Nazis were evil on a level impossible for civilized people to comprehend. The gloves were off. No more trying to follow the Geneva Convention, which expressly banned "murder, mutilation, torture and cruel degrading treatment." The Criminals of the 999th violated every one of those prohibitions in a single night. The gloves were off. In the future Raiding Forces was going to be highly selective about taking prisoners—not in a good way. A cold-blooded resolve was setting in across all agencies represented in the TOC listening to PO3 Nelson's rendition of events.

Lt. Gen. McKoy walked up to the front, "This concludes our initial debrief."

10

THE LEAST OBVIOUS THING

MAJOR THE LADY JANE SEABORN AND BEVERLY Blackwell walked into the Tactical Operations Center. Normally a busy place, this morning it seemed quiet and peaceful. Other than ongoing training missions there were no operations being run out of ABCHQ at present now that what was being called the "Magnificent Mission" had been stood down. When she spotted them, Captain Stephanie Fawcett-Tatum walked over and handed Beverly a TWX.

"I was getting ready to send someone to find you. This message just arrived."

Beverly glanced at the flimsy. "Daddy's sending his plane to take us to London."

Lady Jane said, "When?"

"We're to be picked up at Raiding Forces Headquarters tomorrow."

Lady Jane laughed, "Bronc is not one to wait around. He really swings into action. You better go start packing. We shall be flying to Egypt tonight.

"Stephanie, notify those listed on the manifest I provided you for the trip to London."

Capt. Fawcett-Tatum said, "Wilco."

"Has John been advised?"

"Not yet."

"I shall go give him the good news. We were anticipating our travel to be a few days from now. If there is anything you need before we depart tell us now."

Capt. Fawcett-Tatum said, "I believe Sir Terry and I shall be able to hold down the fort here until you return—enjoy your trip."

Lady Jane went in search of Colonel John Randal. She found him in the Other Ranks mess having a cup of coffee with Petty Officer 3rd Class Joe Nelson. He was informing the Frog that he had been recommended for Officers Candidate School.

"You volunteered for the Navy, OSS, UDT training, Jump School..."

"Sir, I didn't exactly volunteer for Jump School."

"Not a problem. You didn't volunteer for OCS either. You'll enter the class in progress now as soon as you finish your coffee so drink up, Candidate . . . you're already late. Report to Captain Kidd."

"Yes, sir!"

Lady Jane sat down at the table after Candidate Nelson departed. She was laughing. "You certainly did not give the Petty Officer much of a chance to decide, John."

Col. Randal said, "Decide what?"

Lady Jane laughed—his response was typical. "Bronc is sending his plane to fly us to London."

"When?"

"Tomorrow."

Col. Randal would not have wanted her to know travel to London or anywhere else at this time was not something he was looking forward to. Being away from ABC was the last thing he wanted to do with the latest reorganization of Raiding Forces in full swing. He needed to be here—or at least wanted to.

"I'm thinking we should invite Jack," Col. Randal said. "I'd like to show him the Seaborn House operation. That all right with you?"

Lady Jane said, "Absolutely. He needs a vacation. I thought for this trip we would keep our party small. Beverly shall be coming of course, and Brandy and Penelope if they arrive back here in time . . ."

Col. Randal said, “I’d originally asked Mandy to stay and take charge of organizing the MI-5 Counterintelligence program BAD CALL. You have any objection to bringing her as well? I’ll talk to Veronica to see if she can step in and take up her slack.”

“Perfect . . . fun!”

Lady Jane’s green eyes were sparkling so she was happy. That made Col. Randal . . . almost happy.

“Is it true, John, you threatened the Frogs with a knife on board the Hudson?

“Not exactly.”

Second Officer Blyth Colthorp, Vice Admiral Sir Randolph “Razor” Ransom’s WREN, looked in the door, spotted them at the table and came over. “Sir, Admiral Ransom requests your presence in his office at your earliest convenience.”

Col. Randal said, “Inform the Admiral I’m on the way.”

As they were getting up to leave Lady Jane said, “I am so looking forward to this trip.”

Col. Randal lied, “Roger that.”

VICE ADMIRAL SIR RANDOLPH “RAZOR” RANSOM LOOKED UP from the papers he was reading. Colonel John Randal came into his office not sure how this meeting was going to play out. There was a good chance VAdm. Ransom was going to chew him out for leading the drop on Hydra.

The Razor had wanted to conduct the operation to rescue the NAPRW pilot on Sikinos in two phases over a period of days. First the Flying Officer would be rescued. Then while that operation was underway an investigation could be carried out to determine if it was actually Captain Billy Jack Jaxx who had sent the garbled message from Hydra. Or if it was a trap. VAdm. Ransom’s suggestion was the wise and prudent way to attack a complicated high-risk operation with a lot of unknowns. And when the Razor made suggestions he expected them to carry the weight of direct orders.

Disregard them at your peril.

VAdm. Ransom said, "Congratulations are in order, Colonel. While the actual numbers involved were minuscule, you pulled off one of the most complex long-range small-scale Combined Operations I am personally aware of in my thirty-plus-year career. Raiding Forces has developed into quite the 'go anywhere do anything on a moment's notice-capable outfit—impressive."

"Thank you, sir."

"Had you failed I would have preferred court-martial charges."

"I would have expected nothing less, Admiral."

"That is not the reason I asked you here now. You shall be flying out to the UK shortly. My understanding is a meeting with Jane's godfather, Lieutenant Colonel John Henry Bevan, is in order. You two have met before?"

"Yes sir, briefly."

VAdm. Ransom already knew the answer to his question before he asked it. He also knew their meeting had not gone well. Lt. Col. Bevan had informed certain family members—to include himself—that he considered Col. Randal a "lightweight." Members of the Six Hundred, the wealthy aristocratic upper-upper-class families who had controlled England for centuries seldom married outside their circle and virtually never to a foreigner. Unless they were fabulously wealthy or titled in their home country—preferably both.

VAdm. Ransom said, "For reasons that shall become clear, since I am taking sides, we are not having this conversation and if the question should ever arise, I shall deny it."

"Understood, sir."

"Lieutenant Colonel Bevan, the family calls him Johnny the same as Beverly does you, went to Eton—a member of Pop and a graduate of Oxford. He is the son-in-law of the Earl of Lucan and the grandson of the founder of Barclay's Bank. When the last war broke out in 1914 Johnny was commissioned into the Hertfordshire Regiment, rose to the rank of captain, won a Military Cross, and then transferred to Military

Intelligence. He performed analysis which, it was said, he did extraordinarily well. After the war he joined his father's stock brokerage firm and enjoyed enormous financial success.

"When this current war broke out he was recalled to the colors and went back into Military Intelligence. Everyone knows Johnny, but no one knows what Johnny does. He believes intelligence is a business best left to gentlemen. Does not trust anyone outside the military. Takes his role as Jane's godfather seriously.

"Here is what you need to know prior to your meeting. Johnny is going to view you as a bad match for Jane. So here is how to play it. Be strictly professional, officer-to-officer, the way you were with me at first. You hold all the cards—half his age but outrank him by a grade, you have more medals than he does, Hollywood is making a movie about one of your operations and Jane worships you.

"Do not attempt to be Johnny's friend—that would be the kiss of death."

CAPTAIN BILLY JACK JAXX REPORTED TO MAJOR THE LADY Jane Seaborn's suite. He had been summoned. No idea why. The Vulnerable Points Wing NCO on duty said, "Go straight in. Glad you made it back, Captain. Had us worried, sir."

When he walked in Lady Jane came out of the master bedroom carrying a small box. The two enjoyed a special relationship. She was known for taking individuals and units, like OSS OG Det 3 or the Long Range Desert Group, under her wing and then letting them go like baby birds once it was time for them to leave the nest. But Jack Cool was her great favorite. She had no intention of letting him fly away.

To say that he enjoyed their relationship would be a vast understatement. Not only was Lady Jane drop-dead gorgeous, she was a lot of fun to be around. All personality. If he could have picked the perfect aunt she would have been it. The only problem being the hypnotic effect she had on him. It never diminished over time. And there was always the concern when she greeted him by touching him cheek to cheek European style that he might get cut by those razor-sharp cheekbones.

Lady Jane said, “Have you spoken to John?”

Capt. Jaxx said, “Negative, he’s in a closed-door meeting with the Admiral.”

Lady Jane said, “We are flying out for London to be gone ten days to two weeks. John and I would like you to travel with us. Would you care to go?”

“Oh yeah, I’d like that.”

“In that case, sit down . . . we should talk.”

Capt. Jaxx went on full Red Alert. In his experience, when a woman started a conversation with words to that effect what was coming next was not going to be good for him. What could he have possibly done now?

Lady Jane said, “Military etiquette in the UK is quite different than out here in Middle East Command. Much more formal. Mercilessly critical. Rigid in the extreme. And it can be petty. British officers tend to judge each other by details such as the stitching on one’s buttonholes. It may seem silly to you but not any more so than pointy-toed cowboy boots do to us.”

Capt. Jaxx noted she was wearing a pair of crocodile cowgirl boots with her faded blue jeans tucked in as she spoke. Beverly had exerted a Texas-sized influence on the undress uniform wardrobe of the female contingent of Raiding Forces, Lady Jane being no exception. She loved her jeans and boots.

“We shall be staying in my hotel—the Bradford—mingling with high-level British and U.S. officers who stay there when in London. Our bar and restaurant are quiet popular. The Yanks adopt many of our customs. Unfortunately not always the most desirable. They can be even more judgmental than we are.”

Capt. Jaxx wondered where this was going.

Lady Jane said, “You are famed for carrying pin-up photos in your pistol grip. That’s perfectly acceptable out here in Middle East Command but it shall not do in the U.K.—especially at the posh

Bradford. Not if you wish to be taken seriously, and I do so want you to make a good impression, Jack."

"I hear you loud and clear."

Lady Jane said, "John likes smooth pistol grips. He says it is because he grew up learning to shoot with single-action cowboy revolvers," Lady Jane said. "However, for his Christmas present, I intend to surprise him with traditional Turkish Walnut double diamond 1911 grips for his sidearms with extra fine checkering. Thirty lines per inch— too fine to accomplish their task but have the look I desire, with a hand-rubbed oil finish like you would find on a bespoke shotgun or double rifle. If the stocks are still too aggressive for John's taste Westley Richards has agreed to smooth them so the checkering is still visible but not felt to the touch."

Capt. Jaxx said, "Sounds like a great present."

Lady Jane opened the box. Inside were a pair of the grips. They had a dark deadly no-nonsense professional look. "I shall loan a pair of them to you for our trip. You need to make your best appearance, Jack. This is important not just for you but for Raiding Forces, and you will be seen with me. I do not want you to come across as a wild and crazy boy, which you are."

Capt. Jaxx said, "And all this time I thought I was laid-back."

Lady Jane laughed. "Beverly says John is as laid-back as a snake—same applies to you."

Capt. Jaxx said, "I'll take that as a compliment."

Lady Jane said, "Now, as another option if you prefer, I have a pair of the ivory grips with 'Raiding Forces' carved on them in relief that John carried in Abyssinia on his Colt 1911s. The idea was to make him look like a 'Big Shot' to the natives. We could put those on your pistol."

Capt. Jaxx said, "You would do that?"

Lady Jane went back into her bedroom and returned with the ivory stocks she kept in her jewelry box and a small screwdriver. "Let's make the change right now—we shall look like twins. Matching pistol grips."

Capt. Jaxx said, "I can't believe you're doing this."

It was good to be Lady Jane's friend. He had wanted a pair of ivory pistol grips like the ones she commandeered from Col. Randal from the moment he saw them. While Lady Jane had retired her 1911 Colt .38 Super for the Colonel's high capacity 9mm Browning P-35, the High-Power she now wore sported identical ivory Raiding Forces grips—they were going to look like twins.

That was a good thing.

As he was replacing the panels, Lady Jane laughed, "I shall keep your old ones. You must promise to put them back on the moment we return. The girls and I have far too much fun waiting to see who is featured next for you to discontinue the practice."

Capt. Jaxx said, "Deal."

He typically carried his pistol in a radically slanted chest holster that held the weapon sideways for easy access. And also so people could see the photo of his flavor of the week—a Lieutenant General "Geronimo" Joe McKoy design. When Capt. Jaxx slid the handgun back in, he immediately had reservations. What was the Colonel going to say?

"If nobody's told you they love you so far today, Lady Jane—I do."

Lady Jane laughed. "Now there is a price to be paid, Jack. When we arrive in London I shall be taking you to John's tailor for new Class A uniforms. I do not want any trouble. As I explained, the stitching on one's buttonhole is important."

She did not mention John's tailor was also HRH King George VI's personal couturier. Or unknown to him she had shipped one of Capt. Jaxx's Class A uniforms to Pembrook's tailor months ago in anticipation of travel to the UK eventually taking place. Mr. Chauncy, the Master Tailor, would only need to make minor alterations once they arrived in London.

Capt. Jaxx said, "You have my word—Scout's Honor."

Lady Jane was fairly certain Jack Cool had never been a Boy Scout.

CAPTAIN BILLY JACK JAXX WENT DOWNSTAIRS AND headed to the Officers' mess feeling self-conscious about the ivory grips but liking them. He found GG in his office doing whatever it is that mess officers do. Guido "GG" Grazinni, MC, had been captured by Colonel John Randal in Abyssinia. The Italian's dream was to emigrate to the United States after the war and open a restaurant in Hollywood to "serve movie stars." Unknown to him, Major the Lady Jane Seaborn had decided to finance the project.

"Got my stuff, GG?"

"I do, Captain Jack."

The Raiding Forces Mess Officer produced a box. Inside were the thirty-four German 1st Mountain Division stag horn handled knives. GG had polished them until they were gleaming like mirrors.

"You kill all these Nazis?"

"Negative."

"How many?"

"One or two—these look great, GG."

Capt. Jaxx started making the rounds. He gave one of the knives to each OSS OG Det 3 operator. It would have been easier for him to call a formation and hand the knives out all at one time but that would not have been the same as presenting them individually.

Col. Randal spotted him as he was coming out of his meeting with Vice Admiral Sir Randolph "Razor" Ransom. He glanced at the ivory grips but didn't say anything, "What are you up to, Jack?"

"I'm giving one of these 1st Mountain Division knives to each of the Frogs, sir—an *attaboy* for a job well done."

"Looks like you'll have a few left over."

Capt. Jaxx said, "I'm saving those as a Christmas present for Lady Jane to put on her formal dining table. Don't tell her sir. Pretty cool cutting your steak with a dead Nazi's razor-sharp blade."

Col. Randal said, "Yeah, that'll be a nice touch. She only has a ten-place setting. What are you planning to do with the rest?"

"One each goes to General McKoy, Mr. Treywick, Dr. Winthrop, Beverly, Lieutenant Starrett, the Lovats and you, Colonel. I've got enough to go around. I had a couple I'd captured before."

"Give mine to King when he gets here. He did you a service on your Rolex—loves knives."

"Good idea, sir."

"Don't let him know it was my suggestion. This should come from you."

"Thanks, Colonel."

COLONEL JOHN RANDAL WAS IN HIS OFFICE. HE HAD A long list of details he needed to work through before leaving for Raiding Forces Headquarters. Major the Lady Jane Seaborn was not wrong about Major General Sam Houston Blackwell. Bronc was a man of action. When he received word Col. Randal and a small party would be traveling to London, he immediately dispatched his personal, interior-designer-decorated C-47 Dakota to fly them to the U.K., no questions asked. Bronc justified the flight by calling it a "training mission" for the pilots. While it was an extraordinary gesture Col. Randal could have used another day or two to prepare for his time away.

The truth was he did not want to go at all. That might be a mistake. He had been burning the candle at both ends—he knew that. However, Raiding Forces was undergoing a major reorganization due to losses incurred on the Benevento drop and the upcoming departure of the Lancelot Lancers Regiment.

At the same time, Raiding Forces was infusing new outfits like OSS OG Det 3 and 1st Bn KO into the command. And ensuring the attached units like the SBS and the LRDG were brought up to unit standards. Everything happening all at once was creating quite an upheaval and it was placing a strain on his key officers, all the NCOs, and some of the other ranks.

Meanwhile the Raiding Forces squadrons operating on schooners docked along the Turkish coast were keeping the Constant Pressure Concept going. Small patrols were out raiding almost every night. Hit and run.

Col. Randal was willing to concede he could probably use a break but he would have liked to be a part of the reorg—something he enjoyed.

Good commanders pace themselves in order to make sound decisions. He knew that.

Then there was the not-so-minor detail that Lady Jane was excited about the trip. He had never been able to deny her anything—why even bother? It might be time to work on an attitude adjustment. He could do that . . . or at least make a good faith effort to fake it.

Captain Roy Kidd knocked on the door. Col. Randal did not have one of Lady Jane's Royal Marines on a duty desk outside his office to announce visitors because he was so seldom in it. The captain had been flown to ABC from a small remote key where he was conducting training for the SBS officers.

Capt. Kidd said, "You wanted to see me, sir?"

Col. Randal said, "I do, first—on your way out you'll find a submachine gun leaning beside the door. A 9mm Marlin Model 42 called the 'Avenger.' Evaluate it for me. They're OSS issue. Frogs brought 'em but have been using M1 Carbines because they're lighter and handier."

"With pleasure, sir."

The assignment was a double tap: "putting a round peg in a round hole"—a Raiding Forces concept rare in the military—and Rule 4: "Right Man, Right Job." Capt. Kidd was his go-to exotic firearms expert.

Col. Randal said, "Under normal circumstances I'd ask for a detailed rundown on the SBS officer selection and Officers Candidate programs you're supervising. These not being typical times, since I'm due to fly out to England later today, I'll expect a complete report upon my return. That OK with you?"

Capt. Kidd said, "Yes, sir."

"What you're doing with the OCS program and the officer and NCO's Selection Course developing our future leaders is the single most important mission Raiding Forces has going in my opinion—wanted you to know how I feel about it, stud."

"Thank you, sir.

Col. Randal said, "Capt. Jaxx will be traveling with my party, so on top of your other responsibilities you'll be the senior Captain at

ABCHQ. In the event a mission comes up it has to be an absolute top priority to consider it. If in your sole discretion any proposed operation does not meet that standard you have the authority to veto it. I don't want any cowboying or glory missions."

Capt. Kidd said, "Understood, sir." He was thinking the Colonel was the *last* person to be issuing those orders.

In the event something does develop you have to respond to, Lieutenant Starrett will be on call as will Sergeant Major Beckwith—he's available too. You'll have to put together a pickup team of operators, which is not a great formula for success," Col. Randal said. "So don't do it unless there's a clear and present emergency."

Capt. Kidd said, "I understand, sir."

"No long-range multiday missions, period . . ."

Capt. Kidd said, "Colonel, I won't do anything you wouldn't do."

Col. Randal said, "That's not real reassuring, Roy."

VERONICA PAIGE WAS WAITING.

"You wanted to see me, John?"

Colonel John Randal noticed the use of his first name without prompting. *At last*—good. "Congratulations on MI-9's dual rescue of the Northwest African Photographic Reconnaissance Wing pilot and a shipwrecked Captain Jaxx following his escape from the *Perseus*."

Veronica said, "I believe you are the responsible party for the success of those two activities."

Col. Randal said, "That's not how we reported it."

Veronica said, "What did you claim occurred?"

"We stated that acting on information developed by MI-9 Escape under the direction of Mrs. Veronica Paige, assisted by OSS X-2 Counter Espionage Branch officer Beverly Blackwell and supported by Small Raids Inc., Raiding Forces conducted an extraction operation resulting in the simultaneous rescue of both a high-value NAPRW pilot on Sikinos island and one of Raiding Forces' own officers shipwrecked on Hydra Island following the loss of the submarine HMS *Perseus*."

Veronica said, "Why would you not take full credit?"

"You're an equestrian. Didn't anyone ever tell you to never look a gift horse in the mouth? My report gives Raiding Forces credit for the physical going and doing," Col. Randal said. "By stating we were acting on intelligence provided by MI-9 and X-2, both organizations will have maneuvered themselves into a position where they should be able to write their own ticket in the future.

"My guess is General Donovan will be ecstatic about X-2's part. That's a success story he can tell the President . . .'We brought your son Elliott's boy home.'"

Veronica said, "I thought you wickedly clever at Habbaniya but you may have exceeded yourself this time."

Col. Randal said, "As you've probably heard I'm departing for London later today with a small party. Originally I assigned Mandy to take charge of planning our new MI-5 Counterintelligence program—code name BAD CALL—while we're away. But now I'd like her to go with us. Provided, that is, you agree to step in and take over while we're away. She and Beverly can meet with the London headquarters of MI-5 and OSS X-2 to introduce themselves and put faces with names."

"That is a marvelous idea. I shall be delighted to fill in for Mandy," Veronica said. "Now that I have had time to study the situation, in certain instances Counterintelligence and Escape appear to be a natural fit. One thing for sure—we shall not be having any MI-9 vs MI-5 or MI-6 vs SOE competing agendas or inter-agency conflicts here on ABCHQ as is commonly the case at the Puzzle Palace in Cairo. Jim has promised to ensure that does not take place."

Col. Randal said, "Give thought to how you see the two organizations being reconfigured going forward. Once you get your plans fine-tuned, BAD CALL will be off and running. We'll talk when I get back."

Veronica said, "Why do I have the impression you want Mandy along simply because you enjoy her company?"

Col. Randal said, "Because that would be true."

LIEUTENANT COLONEL SIR TERRY "ZORRO" STONE WAS the next person waiting to see Colonel John Randal. He came in but did not close the door. That meant this was a visit, not a meeting.

Lt. Col. Stone said, "Any chance I can fly out to the U.K. with you, old stick?"

Col. Randal said, "Sure."

"I need to get together with Father to coordinate the return of the family regiment to its ancestral home. The Duke has big plans to convert the Lancers to an armored car reconnaissance outfit attached to the Guards Armored Division. Lead the charge ashore on the big day when the invasion takes place."

Col. Randal said, "Must have been a lot of women pining away these last few years who'll be happy to see you back, Sir Zorro."

Lt. Col. Stone said, "Well—there is that. One must keep priorities straight. My time with Father shall be a limited but necessary evil. Too bad Lady Jane shall prevent you from being able to participate in the post-family reunion debauchery."

Col. Randal said, "When can I expect the Lounge Lizards to depart ABC?"

"Sometime after Christmas when transport can be arranged. I was hoping to have the lads home for the holidays but shipping was unavailable," Lt. Col. Stone said. "Not that it shall be a major troop movement by any stretch of the imagination. The Lancers are down to less than a line infantry platoon strength. We always were the smallest Yeomanry Regiment on the Territorial Army list. Now, after all these years of operations with only a trickle of replacements out from the U.K. we are even smaller.

"Mongo would like to remain here in some capacity. I promised him I would speak to you. Will you have a position for him after we depart?"

Col. Randal said, "You bet, Major Farquhar's one of our best officers. We'll form a squadron around him. Have Mongo observe the 11th Parachute Battalion and the 1st Battalion, King's Own troops going through selection for men he'd like to recruit. He can have his choice."

Lt. Col. Stone said, “Percy is up to speed on his new duties. My traveling with you shall not pose any problems on that score—time for him to take the reins.”

Col. Randal said, “As long as you’ve changed the lock on the explosives locker and not given Pyro a key, we’re good.”

Lt. Col. Stone said, “All right, then . . .”

He got up to go. This was a painful conversation between two best friends. Lt. Col. Stone did not want to leave Raiding Forces but he had family obligations. Col. Randal did not want to see him go but he had a command to run.

In the army nothing is forever.

JAMES “BALDIE” TAYLOR WAS IN NEXT. AN MI-6 OFFICER masquerading as an SOE local Major General who was serving as the liaison between OSS and both of the British intelligence agencies—at times it was difficult to know whose interest he was representing. He and Colonel John Randal had a long—as measured in war years—relationship and while they might not be considered close friends they had developed a mutual respect for each other.

Lieutenant General “Geronimo” Joe McKoy had advised Col. Randal, “Keep your eye on Jim—he’s got the makin’s of a’ snake in the grass.” That said, often when a mission was imminent Baldie would show up with a weapon and go to war with Raiding Forces. He was a complicated individual.

Jim said, “Mind if I straphang your flight to London?”

Col. Randal said, “Not at all. Any special reason?”

“Donovan is there to link up with us in London. He knows you are meeting with Lady Jane’s godfather, Colonel Bevan. And he is aware you plan to consult with Commander Fleming about his Red Indian/40 Commando project. Wild Bill is highly interested in the outcome of both those meetings.

"While I am not invited to either of those events," Jim said, "I need to be available to shepherd him through the murky underworld of British intelligence if questions arise afterward—as they always do."

Col. Randal said, "Tell Jane."

Jim said, "Rocky needs to travel with us as well."

"Why might that be?"

"That's classified."

KING AND CAPTAIN PAMALA PLUM-MARTIN CAME INTO the TOC. Colonel John Randal spotted them through the big plate glass window and waved them to come over. The two were dressed for their role as a wealthy power couple and looked every inch the part. They made a handsome pair.

Col. Randal said, "You can brief me on Morocco later. Time is short. Bronc jumped the gun and sent his private plane without first checking to see what our itinerary was. We're flying out to RFHQ in a few hours—that is, if you still want to go."

Capt. Plum-Martin laughed. "Surely you jest, love. King and I cut our trip short to make the London trip. We would not miss it for the world."

Col. Randal had seldom seen the beautiful snow blonde so animated. Possibly it was the idea of being home for the Christmas season—at least part of it. The plan was to return to ABC before December 25.

The Merc was impossible to read. He might be humoring Capt. Plum-Martin. Some of that was going around.

After they left, Major Butch "Headhunter" Hoolihan came in.

Col. Randal said, "What are you doing here, Butch?"

Maj. Hoolihan said, "I turned my Red Indian Marines over to Mad Dog, sir. The OSS OG Det 3 have sustained so many jump-related injuries they are not able to continue their parachute training for at least another two weeks.

"I spoke with Jack about them. He tells me they are excellent material—appalling patrolling technique. We decided to have the Red Indian Marines take their place in Jump School while I organize a program of patrolling here on ABC for the Frogmen. It shall be crawl, walk, run, initially focusing on technique. Gentle enough, at least at the start, to allow for them to recover from their jump injuries. Then we will ramp it up as they get back to a hundred percent."

Col. Randal said, "Good. Correct their deficiencies, Butch, and you're going to have a great outfit once you get organized."

"What do you think I should name it, Colonel?"

"How about Red Frog Squadron—Lady Jane suggested it?"

He did not say it but Col. Randal was thinking "Right Man, Right Job." With quality officers like the "Headhunter" who bought into the Raiding Forces culture of "teamwork, teamwork, teamwork," things got done the way they should without having to be told what, when or where.

Maj. Hoolihan said, "In that case Red Frogs it is sir."

While the reorganization currently underway might appear chaotic to an outsider, Col. Randal was comfortable things were in good hands and progressing according to plan. He did not have a profusion of officers and NCOs due to the attrition rate in leadership slots always being high with few replacements. The ones he did have were committed, battle-tested professionals. The best in the business.

They could be counted on. All Col. Randal had to do was provide a clear-cut assignment/mission statement. Then step back and let them get the job done. Not all commanders had the luxury of subordinates qualified to take charge or were unwilling to relinquish authority.

There were a couple of reasons it was smart to let subordinates run their own show whenever possible. It demonstrates trust between the commander and his officers and NCOs. And it cultivates independent initiative within the unit.

Not many Commanding Officers understand that.

NEWLY PROMOTED LIEUTENANT JUNIOR GRADE WESTLY Slade knocked on the door after Major Butch Hoolihan departed.

"You wanted to see me, Colonel?"

"Roger that. You're expected in Achnacarry, Scotland, no later than zero six hundred hours Monday. Report to Colonel Vaughn. You're slated to take the Commando Basic Training Course—polish your patrolling skills. I expect you to be the Honor Grad."

LTJG Slade said, "Anything else I need to know, Colonel?"

"Like what?"

"How do I get there, sir?"

"You figure it out, Lieutenant."

As soon as the confused Frog departed, Col. Randal picked up the phone. "Stephanie, locate Captain Jaxx and have him step in my office."

"He's right here, John. I shall send him over."

Captain Billy Jack Jaxx arrived in seconds. "You needed to see me, sir?"

Col. Randal said, "Lieutenant Slade has been ordered to report to Achnacarry for Commando training. You might want to mention to him how Raiding Forces personnel "Frogspawns" transportation when they need to travel. No rush, keep it casual. Just make it happen before we fly out. And Jack . . . we never had this conversation."

"Can do, sir."

Col. Randal said, "You up for this expedition?"

Capt. Jaxx said, "Can't wait. Zorro's taking a few of us to the Windmill Theatre when we hit town, sir."

"What's that?"

"Don't know for sure. It's an exclusive 'men only' nightclub with a floor show he claims I'll never forget. I'm not supposed to tell Lady Jane, sir so don't mention it. Colonel Stone's father keeps box seats there."

Col. Randal said, "Really."

BEVERLY POPPED IN AS CAPTAIN BILLY JACK JAXX WAS leaving. She sat on the edge of Col. Randal's desk and crossed her legs. "Have your bags packed, Johnny?"

Col. Randal said, "Not yet. I've had a lot of loose ends that needed to be taken care of first."

"Nervous about meeting the godfather-in-law?"

"Don't think there's any such thing, Beverly. I might be . . . maybe."

Beverly said, "Daddy's going to join us at the Bradford. We're toast as far as keeping him from finally meeting Brandy. She and Penelope have both signaled they'll be arriving in time to fly out this evening."

Col. Randal said, "You gave it your best shot."

"Lady Jane's planning a day to ride with the Beaufort Hunt," Beverly said. "Any chance of you joining us?"

Col. Randal said, "I have no idea what my schedule is."

"Would you say you owe me one for Hydra?"

"I would."

"OK then—pay me back by standing down. Just this one time, Johnny. Relax, enjoy—we'll have a great time."

"I'm working on that."

CAPTAIN STEPHANIE FAWCETT-TATUM ARRIVED IN Colonel John Randal's office. She shut the door. The tall glamorous brunette was a longtime friend—the third handpicked member when Major the Lady Jane Seaborn originally formed her Royal Marine detachment. She and Lady Jane had nursed Col. Randal back to health after he had been shot, which meant having seen him at his weakest. She knew him about as well as anyone in Raiding Forces.

"Since you shall be flying out this evening it is imperative to dictate your After Action Report while it is still fresh on your mind, John."

After Action Reports were a staple of Raiding Forces. From time to time some higher headquarters requested a copy after a mission it had an interest in—usually meaning one they could take some credit for.

Generally, Col. Randal read the ones his subordinate commanders wrote and made the reports required reading for all officers and NCOs and optional reading for the troops. The idea was to analyze the run-up to a mission's phase, the actions on the objective and the withdrawal for "Lessons Learned"—meaning what worked and what did not. Every individual who participated gave a verbal report of their actions from start to end of mission to one of the Royal Marines. They took down the reports in shorthand, typed them up, made copies, had them collated, then signed for. After they had been read, the documents were turned in and locked away in the unit safe stamped SECRET. The reports were not for public consumption. And they were not something that should fall into enemy hands.

Col. Randal said, "Is this absolutely . . ."

Capt. Fawcett-Tatum said, "It is, John. You know our protocol. It is how we do things."

Col. Randal said, "Intelligence indicating a NAPRW pilot was down on . . ."

LADY JANE STUCK HER HEAD IN THE DOOR AND FLASHED one of her patented heart attack smiles. "If you keep inviting people, I shall have to ask Bronc to send another plane. This trip is already tremendously fun and we have not even departed yet."

Col. Randal was beginning to come around to the idea she might be right—everyone else had.

COLONEL JOHN RANDAL WAS SPEAKING TO THE assembled OSS OG Det 3 and attached parties, taking them through an exacting step-by-step debriefing. Everyone who could be there was in the TOC to hear him. Major the Lady Jane Seaborn was on hand to make sure he would walk out with her and board the Walrus as soon as he finished. Their bags were already onboard.

The performance was a tour de force. Over the last two years the Frogs had sat through countless canned training debriefings. Due to its complexity and scope they had never heard the equal of this one. The delivery was simple, to the point and brutally honest. Col. Randal did not rush. The presentation took an hour.

"Regarding the drop on Hydra's built-up area," Col. Randal said, "the purpose of the exercise was to do the *least* obvious thing. By doing so we were able to take advantage of the shock effect of an airborne forced entry assault to create maximum disruption among the German defenders. By dropping directly on the village we amplified the element of surprise—the speed and violence of action served as a force multiplier."

The explanation was so simple and tactically sound that people in the TOC who had questioned Col. Randal's selection of the DZ were left wondering why they had failed to recognize his choice for what it was. A brilliant, if risky, tactical maneuver.

Each member of the Sikinos and Hydra teams—to include Lieutenant General "Geronimo" Joe McKoy, Doctor Layton Winthrop, the Lovat Scouts, Beverly Blackwell, Lieutenant Westly Slade, Lieutenant Ted Hamilton, the High-Speed Launch officers, the Life Boat Service Men and Waldo Treywick—were present to discuss their individual roles. Criticism was invited and given. There are no perfect missions.

Finally Col. Randal said, "This concludes my debriefing. Now there are a couple of administrative details I'd like to take care of before I fly out."

Lady Jane was impatient to go. She gave him a look. Which he ignored.

Col. Randal said, "I see Major Hoolihan has joined us. Effective immediately OSS OG Det 3 is going to be assigned to him. He'll be organizing a highly classified Raiding Forces special operations unit working for British Naval Intelligence and the Office of Strategic

Services. You'll be teamed up with a handpicked troop of Royal Marines he's currently training.

"Better shape up—Major Hoolihan's not called the 'Headhunter' without cause."

The Frogs all turned in their chairs to check out their new boss. They had heard of him. A Raiding Forces legend.

Col. Randal said, "One last thing, 'Operational Swimmer' doesn't cut it for me. You men deserve a more fitting designator. In the British Army, titles in Commando type units are considered important enough to be called ranks. Effective immediately members of OSS OG Det 3's new 'rank' will be 'Special Warfare Operator.'"

Judging from the Frogs' reaction, they liked the new name.

But not all of them. Two sought out LTJG. Slade immediately after the debriefing. They requested to be returned to their unit—as was their right. The men would be sent back to the U.S. space available surface shipping without prejudice. At least on paper.

Vice Admiral Sir Randolph "Razor" Ransom had them off the island aboard a locally chartered caique within the hour, working the first leg of their passage home as deckhands bound for the U.S. Naval base at Casablanca—over 2,000 nautical miles distance. A voyage that would take at least a month, weather permitting. The two Frogs departed ABC after having been relieved of their brand new Jump Wings. Clearly, "without prejudice" did not mean with no hard feelings. The Razor had no sympathy for men who could not hack it. There was a war on.

With two men KIA, two quitters and everyone else suffering one injury or another, the Special Warfare Operators(SWO) of OSS OG Det 3 had suffered a high casualty/attrition rate for less than a single week's duty in Raiding Forces—a unit they were quick to concede was not for "the weak or faint of heart" as had been advertised.

The Frogs were good with that—still it was hard to lose men they had trained with for two years.

MAJOR "PYRO" PERCY STIRLING WALKED COLONEL JOHN Randal and Major the Lady Jane Seaborn down to the Walrus that would be transporting the party to RFHQ for the night to marshal for the trip to London. Happy was trotting along with them but for once he was not living up to his name. The animal sensed he was being left behind.

Col. Randal said, "You're in command, Major. I want you to continue the training. Don't take on any direct action missions unless you absolutely have no other option. As we discussed, I've authorized Captain Kidd to veto any operation unless it's an absolute emergency—he'd be the troop commander in the event something comes up that can't be put off until I return."

Maj. Stirling said, "I am quite clear on that, sir."

"I didn't mean to throw you into the fire so quick, Percy. You know I wouldn't do it if I wasn't confident you're up to the job, one hundred percent," Col. Randal said. "Stephanie endorsed you, so do what I do—stay out of the way and let her run the show as much as possible."

"My plan down to the ground, Colonel.

Lady Jane said, "You and Stephanie take care of my dog."

Then they boarded. Captain Pamala Plum-Martin and Beverly Blackwell were the pilots tonight. The plane began its takeoff run as soon as Col. Randal and Lady Jane were in their seats. As the Walrus flew through the rapidly darkening sky, Lady Jane was resting her arm across Col. Randal's shoulders with her scarlet nails splayed on his chest—a very possessive posture which was how she was feeling. Col. Randal started to unwind.

As always she took away his pain.

Waldo Treywick was saying, "Me and P. J. Pretorious got word there was a tribe a' man-eatin' gorillas..."

Lieutenant Ted Hamilton aka "The Great Teddy" was saying—*what was he doing on the plane?*—"A magician is an actor dressed up in a tuxedo attempting to deceive his audience. I am an illusionist . . ."

Lieutenant General "Geronimo" Joe McKoy was saying, "Now your average GI can't hit the side of a phone booth with the standard issue Government Model forty-five if he's standing inside . . ."

James “Baldie” Taylor was saying, “After my third divorce my attorney advised me, ‘Next time why not just pick out some woman who hates your guts and buy her a house’ . . .”

Beverly, who had come back from the flight deck to visit, was saying, “Daddy warned me about cowboys and Lone Star beer but he didn’t say anything about women and cocaine . . .”

Col. Randal dozed off. All was right with his world—for now.

~ ~

THE MISSION CONTINUES IN

–Military Deception –

BOOK XVIII IN THE RAIDING FORCES SERIES

COMING SOON

~ ~

The Raiding Forces series continues all the way to VE Day.
To be on our notification list for the next book, contact
phil@philward.com

ABBREVIATIONS
ORDERS & AWARDS

Bt	Baronet
CB	Companion of the Bath
CMG	Companion of the Order of St. Michael & St. George
DCM	Distinguished Conduct Medal
DFC	Distinguished Flying Cross (Royal Air Force)
DSC	Distinguished Service Cross (Royal Navy)
DSM	Distinguished Service Medal
DSO	Distinguished Service Order
GC	George Cross
GCB	Grand Cross in the Order of the Bath
GM	George Medal
KBE	Knight Commander of the Most Excellent Order of the British Empire
KCVO	Knight Commander of the Royal Victorian Order
LG	Lady Companion of the Order of the Garter
MC	Military Cross
MM	Military Medal
MVO	Member of the Royal Victorian Order
OBE	Order of the British Empire
SS	Silver Star Medal (U.S. Armed Forces)
VC	Victoria Cross

LIST OF CHARACTERS

Alexandra (Mandy) Paige, OBE, RM
Artemus, Greek contact (SOE)
Beverly Blackwell, SS
Brig. Gen. Raymond J. "R. J." Maunsell
Brig. Gen. William "Wild Bill" Donovan
Capt. Billy Jack Jaxx, MC
Capt. Clint Hays
Capt. Dan Bonham
Capt. Jake Novak aka Jake the Snake
Captain Jeffery Tall-Castle, MC, OBE
Capt. Karen Montgomery, RM
Capt. Pamala Plum-Martin, DSO, OBE, DFC RM
Capt. Preston Butterfield III
Capt. Richard Holden
Capt. Roy "Mad Dog" Reupart, MC
Capt. Stephanie Fawcett-Tatum, RM
Cdr./Gen. Frank Polanski
Cdr. Ian Fleming, RNVR
Col. Elliott Roosevelt
Col. John Randal, DSO, OBE, DSC, MC
Lt. Colonel John Henry Bevan, MC
Cpl. Raymond Davenport
Cpl. Tom "Murph the Surf" Murphy
Dr. Stephen Milam
Flanigan
FO Bash Potgieter
Guido "GG" Grazinni, MC
Happy
Henry Greathead, Life Boat Service Man
King
Lana Turner
LCpl. Ray "Tank" Karlsson
Lovat Scout Lionel Fenwick
Lovat Scout Munro Ferguson
Lt. Cdr. Edward "Ted" Nicolay
Lt. Coco Lovejoy, Wren
Lt. Gen. "Geronimo" Joe McKoy
Lt. Jeffery Featherstonehaugh, RNVR
Lt. Malcolm North
Lt. Ted Hamilton, OBE aka The Great Teddy
Lt. Violet Pelkington
LtJg. Westly Slade
Maj. Butch "Headhunter" Hoolihan, DSO, MC, MM, RM
Maj. Gen. Sam Houston "Bronc" Blackwell
Maj. the Lady Jane Seaborn, LG, OBE, RM
MSgt. Mack Beckwith
PFC Bray Landreth
PFC David Schwartz
PFC Norvel "Horn Dog" Hansen
PFC Wally Malinowski
PO3 Joe Nelson
Pvt. Nathan Boatwright
Rikke (Rocky) Runborg
Rita Hayworth
S/Lt. Peter Grey
Second Officer Blyth Colthorp (2/O), Wren
Stanley Bleecker , Life Boat Service Man
VAdm. Sir Randolph "Razor" Ransom, VC, KCB, DSO, OBE, DSC, RN
Veronica Paige, OBE
Waldo Treywick
W/Cdr Paddy Wilcox
Xanthos

ACRONYMS

AA	Antiaircraft
ABC	Advanced Base Castelrozzo
ABCHQ	Advanced Base Castelrozzo Headquarters
AO	Area of Operation
ATO	Aegean Theatre of Operation
BMNT	Begin Morning Nautical Twilight
Bn	Battalion
BPBC	British Power Boat Company
CI	Counterintelligence
CO	Commanding Officer
CP	Command Post
CPC	Constant Pressure Concept
Det 3	Detachment 3
DSEA	Davis Submerged Escape Apparatus
DZ	Drop Zone
DZST	Drop Zone Support Team
DZSTO	Drop Zone Officer-in-Charge
E&E	Escape and Evade
EM	Enlisted Man/Men
ETO	European Theatre of Operations
FANY	Field Auxiliary Nursing Yeomanry
GSS	Greek Sacred Squadron
HMG	Heavy Machine Gun
HSL	High-Speed Launch
IP	Initial Point
KO	King's Own
KIA	Killed in Action
LBSM	Life Boat Service Men
LCR(S)	Landing Craft Rubber (Small)
LCS	London Controlling Station
LD	Line of Departure
LRDG	Long Range Desert Group
LSF	Levant Schooner Flotilla
MAS	*Motoscafo armato silurante*, torpedo armed motorboat (Italian)
MEHQ	Middle East Command Headquarters
METS	Middle East Training School
MGB	Motor Gunboat
MI	Military Intelligence
MIA	Missing in Action
ML	Motor Launch
MND	Marinen-achrichtendienst (German Naval Intelligence Service)
MO	Medical Officer
MTB	Motor Torpedo Boat
NAPRW	Northwest African Photographic Reconnaissance Wing
NCOIC	Noncommissioned Officer in Charge
No.4 METS	-No. 4 Middle East Training School
OCS	Infantry Officers Candidate School
OG	Operational Group
OKM	Kriegsmarine Naval Intelligence Department 6
OSS	Office of Strategic Services aka the Outfit
X-2	Counter Espionage Branch

OWI	Office of War Information
PAX	Passengers
PB	Parachute Battalion
PC	Police Chief
PETN	Pentaerythritol Tetranitrate
PLF	Parachute Landing Fall
QRS	Quick Release System
QRT	Quick Reaction Team
RAF	Royal Air Force
RAMC	Royal Army Medical Corps
R&R	Rest and Recreation
RFHQ	Raiding Forces Headquarters
RON	Remain Overnight Position
RTO	Radio Telephone Operator
RTU	Returned to Unit
S-3	Operations Officer
SBS	Small Boat Squadron
SIGNIT	Signals Intelligence
SIS	Secret Intelligence Service (British)
S2	Intelligence Officer
SMG	Submachine Gun
SOE	Special Operations Executive
SOG	Small Operations Group
SOP	Standard Operating Procedure
SRI	Small Raids Incorporated
TDY	Temporary Duty
TOC	Tactical Operations Center
TO&E	Table of Organization and Equipment
TTC	Troop Transport Command
TWX	Teletypewriter Exchange Service

ABOUT THE AUTHOR

Phil Ward is a decorated combat veteran commissioned at age nineteen. A former instructor at the Army Ranger School, he has had a lifelong interest in small unit tactics and special operations. He lives in Texas on a mountain overlooking Lake Austin.

~~~

OTHER BOOKS IN THE RAIDING FORCES SERIES

Those Who Dare

Dead Eagles

Blood Wings

Roman Candle

Guerrilla Command

Necessary Force

Desert Patrol

Private Army

Africa 1941

The Sharp End

Raiding Rommel

Strategic Services

Tip of the Sword

Always So Few

The War That Never Was

Economy of Force
~~~

www.ingramcontent.com/pod-product-compliance
Lightning Source LLC
Chambersburg PA
CBHW021624030826
48979CB00036B/2038/J
* 9 7 9 8 9 9 2 0 6 4 5 0 6 *